#1 JOP AND BLIP WANNA KNOW

CAN YOU HEAR A PENGUIN FART ON MARS?

AND OTHER EXCELLENT QUESTIONS

JIM BENTON

#1

JOP AND BLIP WANNA KNOW

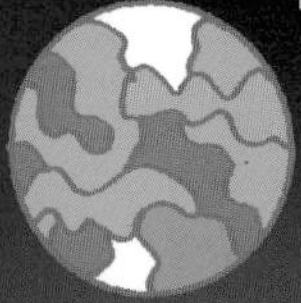

HARPER alley
An Imprint of HarperCollinsPublishers

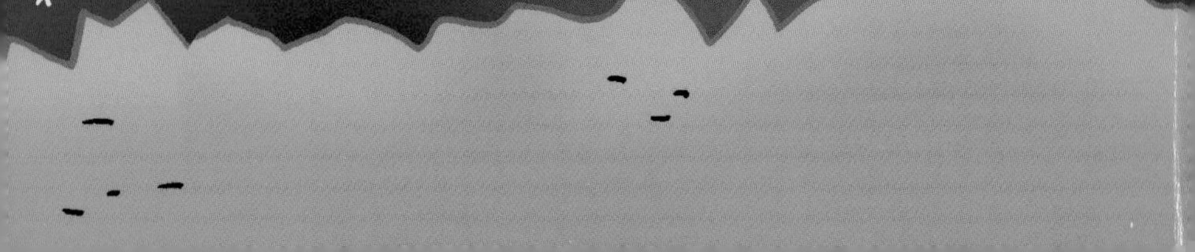

TO ALL THE SCIENTISTS WHO WORK SO HARD TO HELP US.

HARPERALLEY IS AN IMPRINT OF HARPERCOLLINS PUBLISHERS.

JOP AND BLIP WANNA KNOW #1: CAN YOU HEAR A PENGUIN FART ON MARS?
COPYRIGHT © 2021 BY JIM BENTON
ALL RIGHTS RESERVED. PRINTED IN BOSNIA AND HERZEGOVINA.
NO PART OF THIS BOOK MAY BE USED OR REPRODUCED IN ANY MANNER WHATSOEVER WITHOUT
WRITTEN PERMISSION EXCEPT IN THE CASE OF BRIEF QUOTATIONS EMBODIED IN CRITICAL ARTICLES
AND REVIEWS. FOR INFORMATION ADDRESS HARPERCOLLINS CHILDREN'S BOOKS, A DIVISION OF
HARPERCOLLINS PUBLISHERS, 195 BROADWAY, NEW YORK, NY 10007.
WWW.HARPERALLEY.COM

LIBRARY OF CONGRESS CONTROL NUMBER: 2020947267
ISBN 978-0-06-297292-7 — ISBN 978-0-06-297293-4 (PBK.)

DRAWN WITH A FLAIR PEN ON CHEAP PAPER AND COLORED IN PHOTOSHOP.
TYPOGRAPHY BY JIM BENTON AND ERICA DE CHAVEZ
21 22 23 24 25   GPS   10 9 8 7 6 5 4 3 2 1
❖
FIRST EDITION

TABLE OF CONTENTS

THE PENGUIN PUZZLE

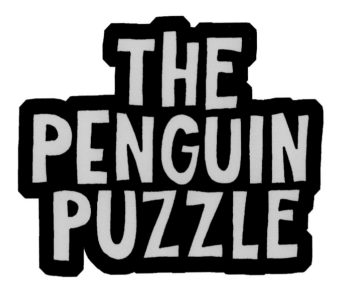

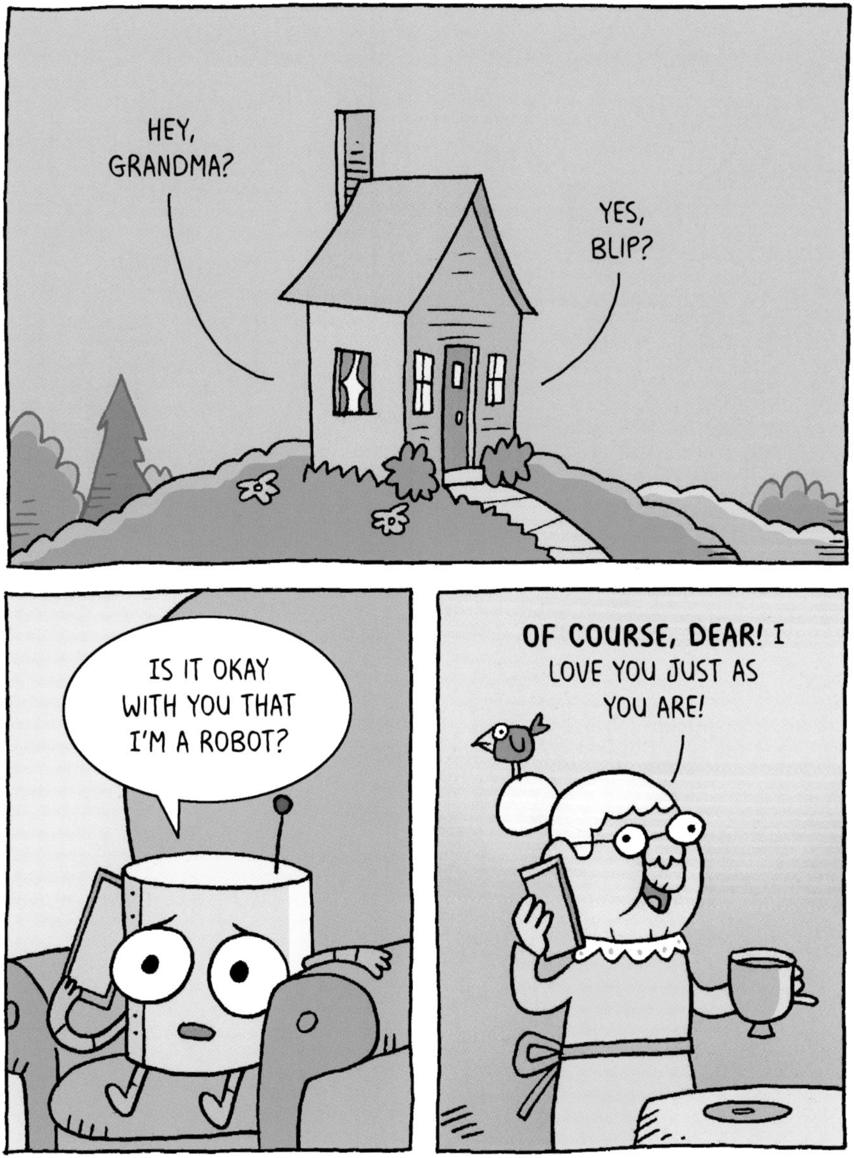
HEY, GRANDMA?
YES, BLIP?
IS IT OKAY WITH YOU THAT I'M A ROBOT?
OF COURSE, DEAR! I LOVE YOU JUST AS YOU ARE!

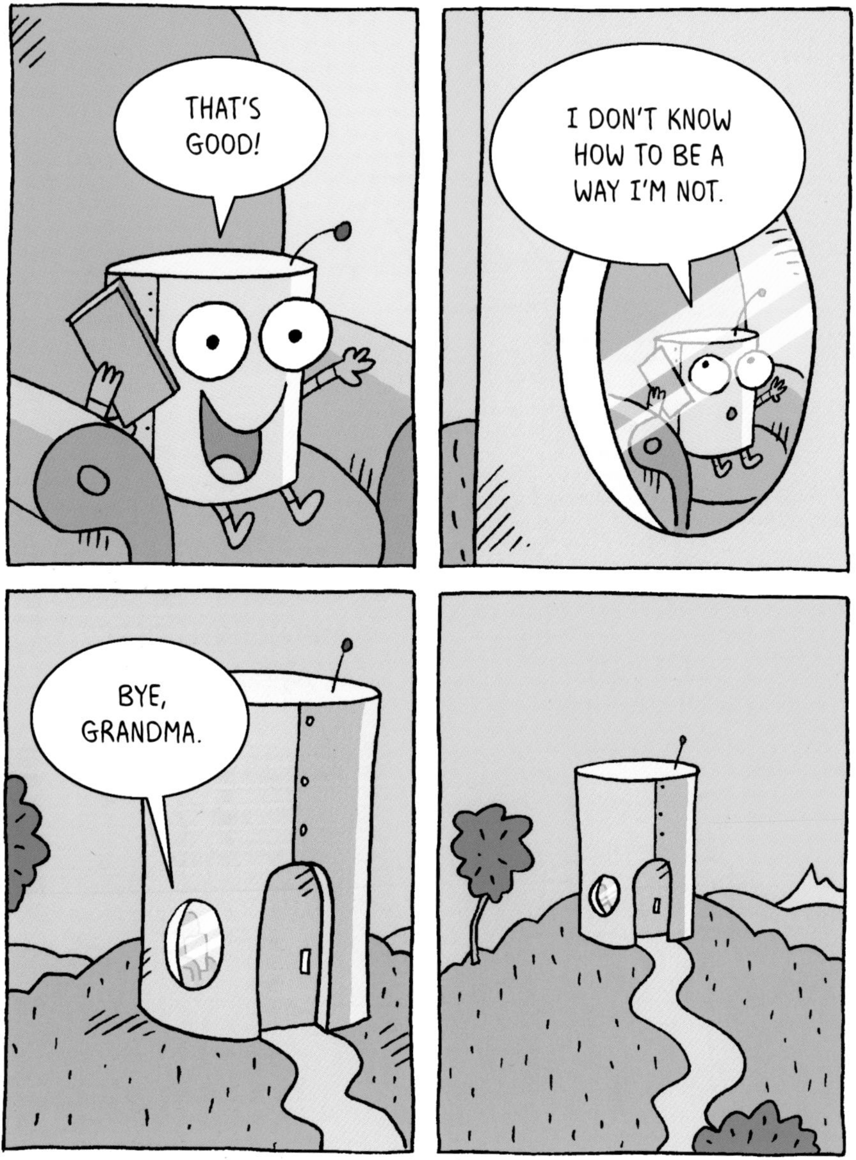
THAT'S GOOD!
I DON'T KNOW HOW TO BE A WAY I'M NOT.
BYE, GRANDMA.

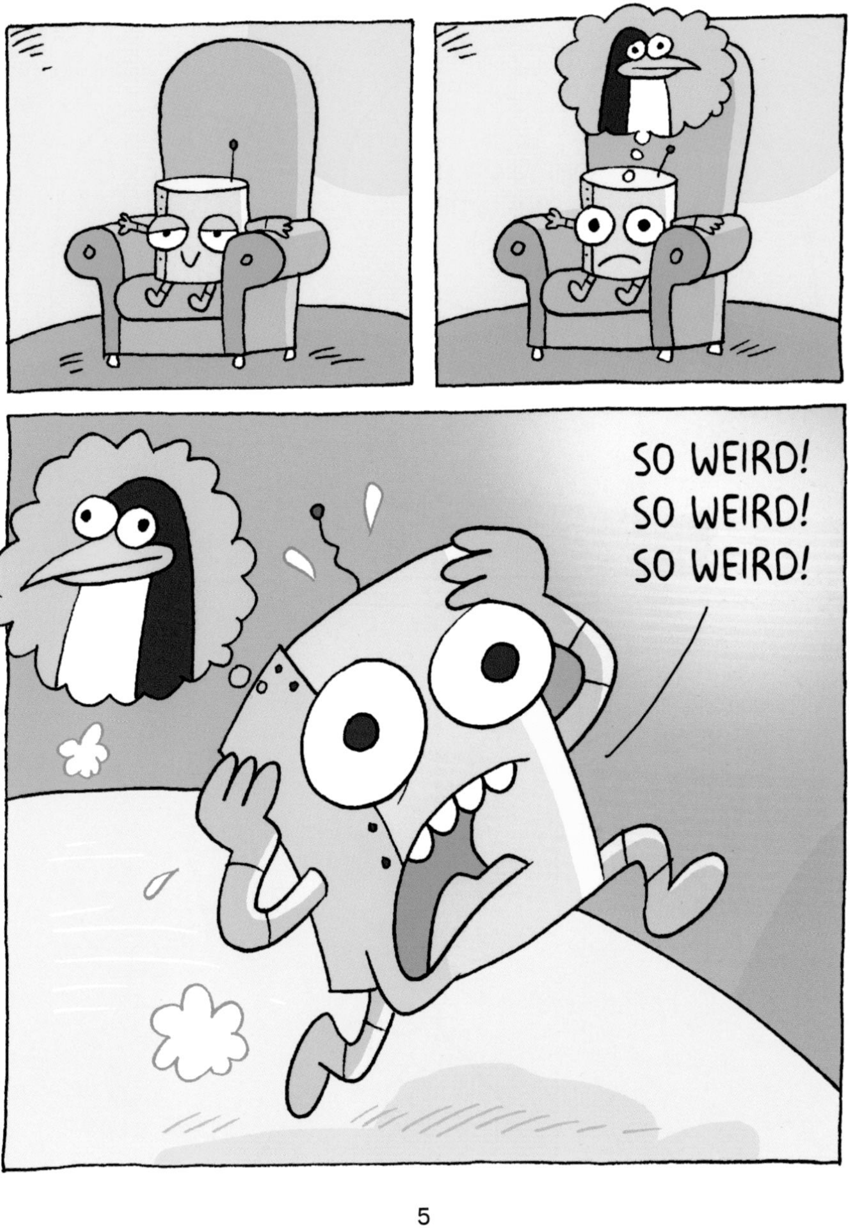
SO WEIRD!
SO WEIRD!
SO WEIRD!

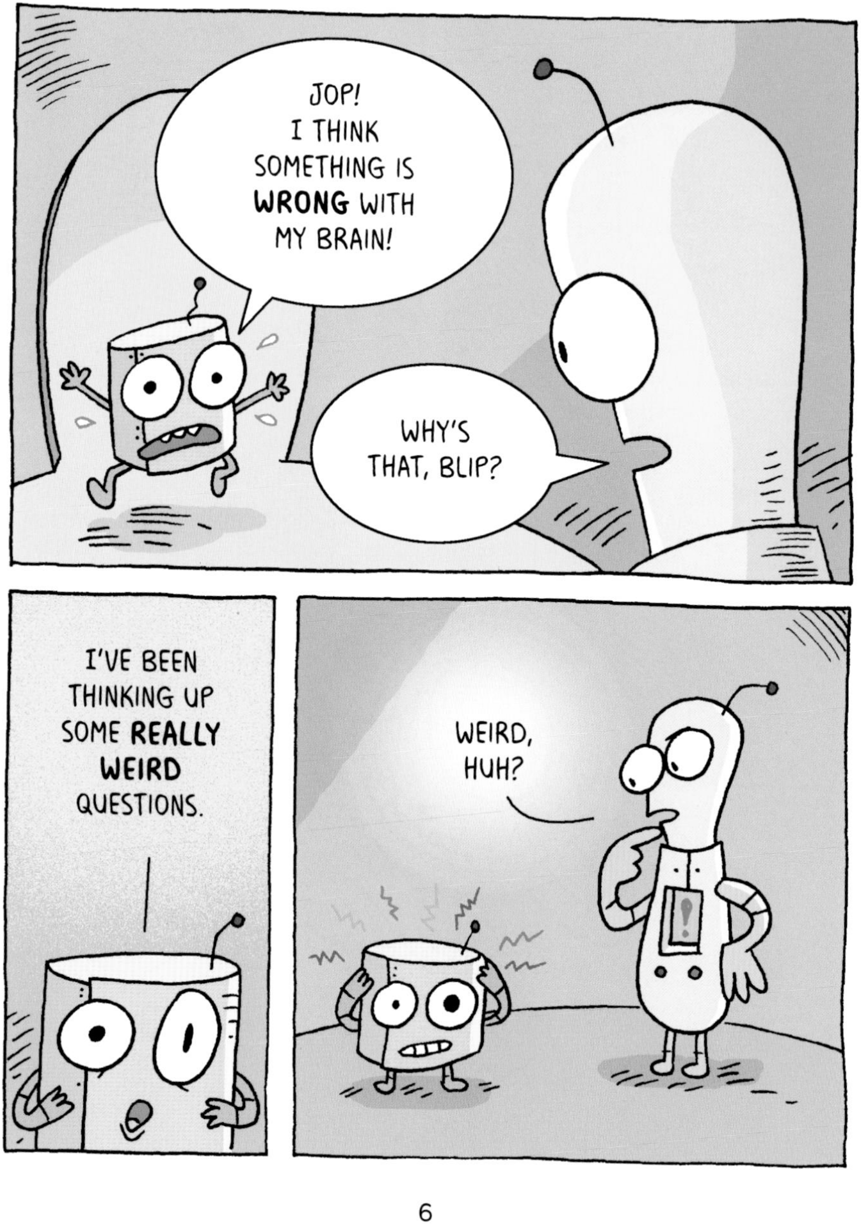
JOP!
I THINK SOMETHING IS **WRONG** WITH MY BRAIN!
WHY'S THAT, BLIP?
I'VE BEEN THINKING UP SOME **REALLY WEIRD** QUESTIONS.
WEIRD, HUH?

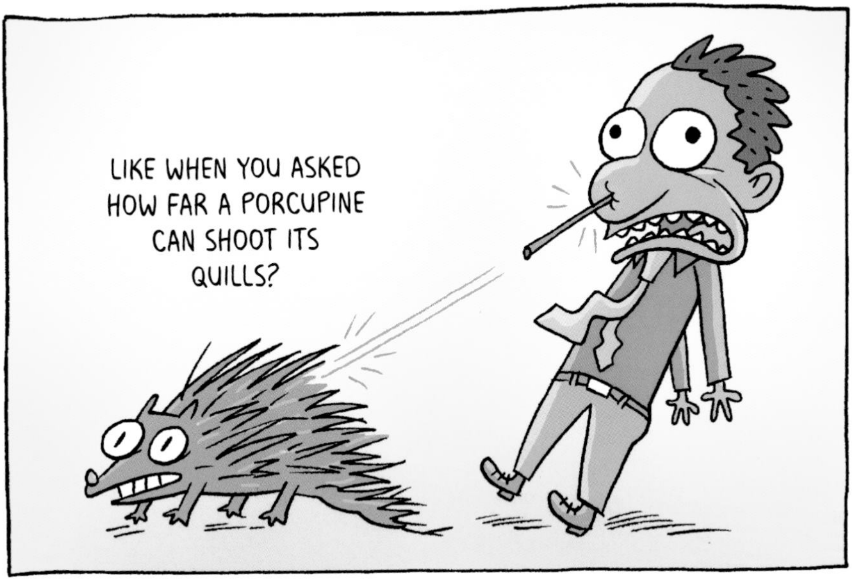
LIKE WHEN YOU ASKED HOW FAR A PORCUPINE CAN SHOOT ITS QUILLS?

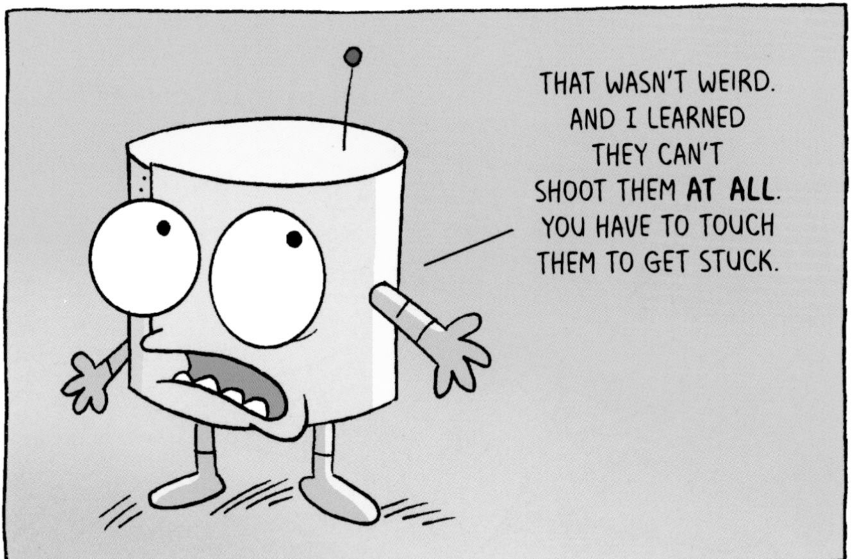
THAT WASN'T WEIRD. AND I LEARNED THEY CAN'T SHOOT THEM **AT ALL**. YOU HAVE TO TOUCH THEM TO GET STUCK.

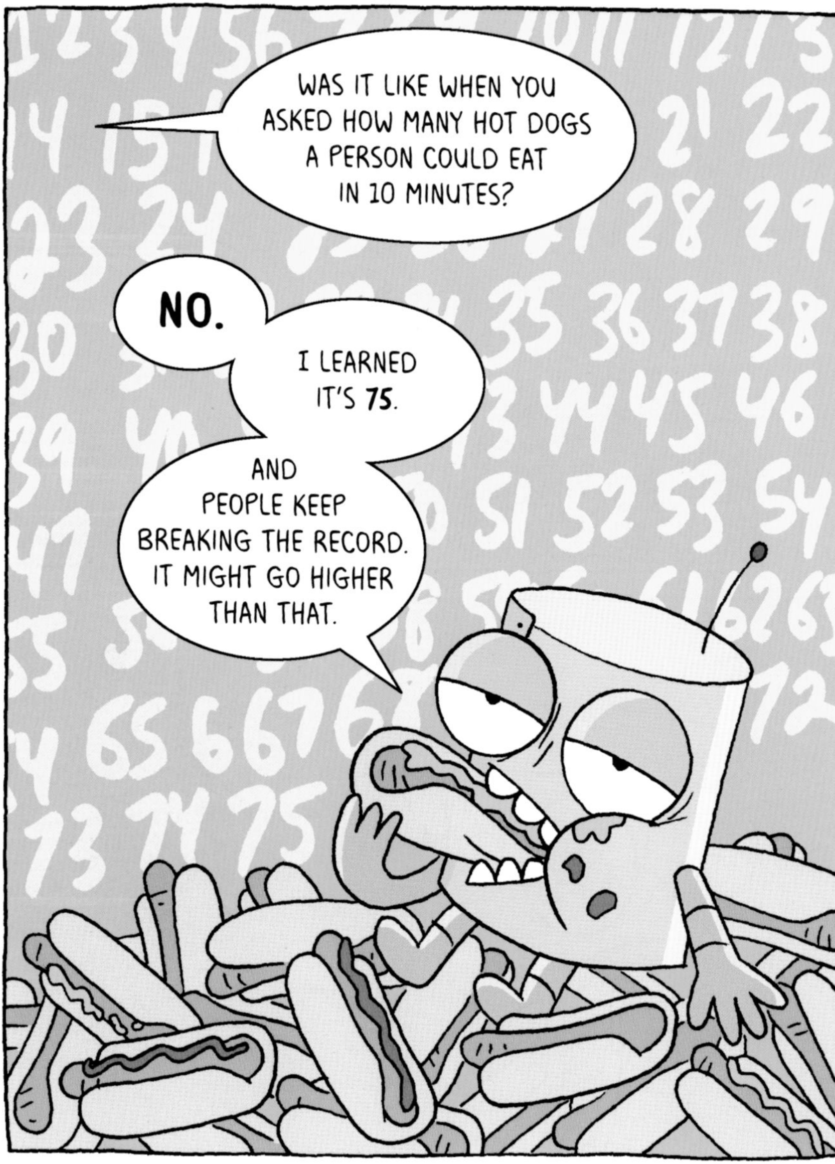
WAS IT LIKE WHEN YOU ASKED HOW MANY HOT DOGS A PERSON COULD EAT IN 10 MINUTES?
NO.
I LEARNED IT'S 75.
AND PEOPLE KEEP BREAKING THE RECORD. IT MIGHT GO HIGHER THAN THAT.

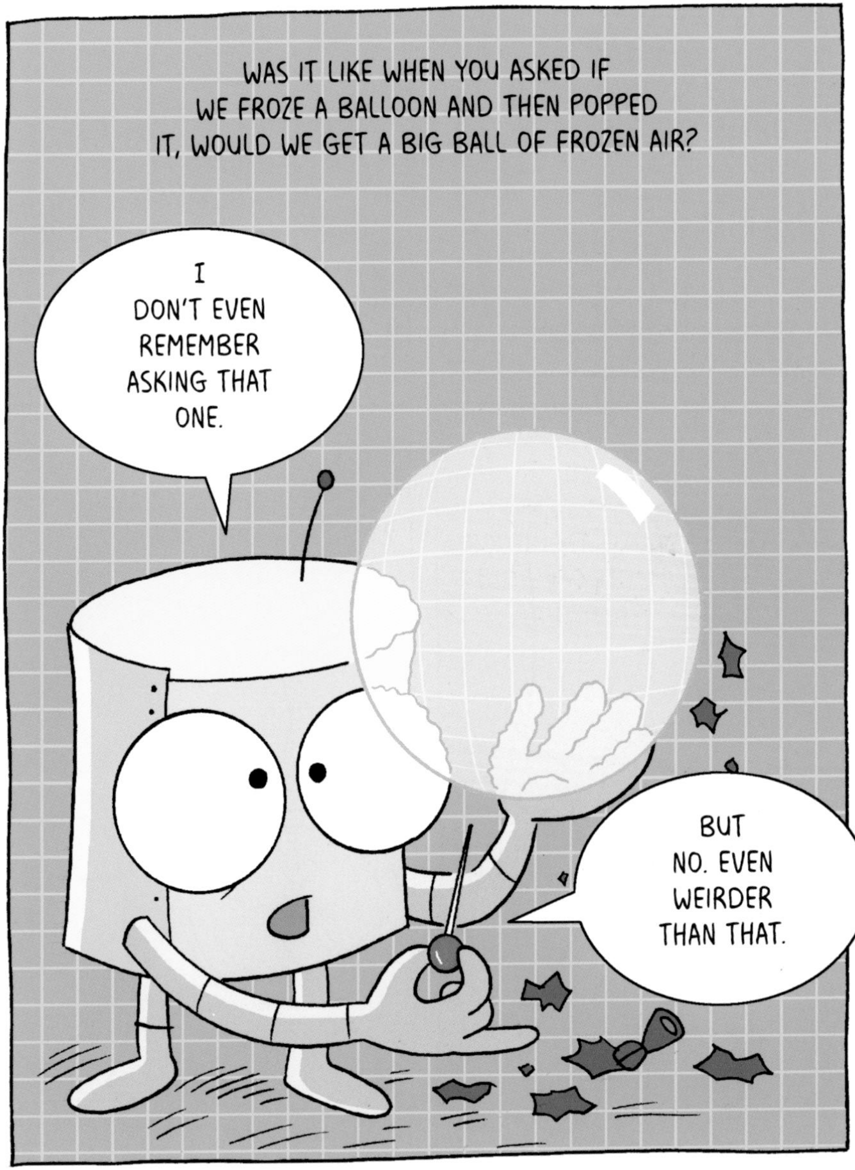

WAS IT LIKE WHEN YOU ASKED IF WE FROZE A BALLOON AND THEN POPPED IT, WOULD WE GET A BIG BALL OF FROZEN AIR?
I DON'T EVEN REMEMBER ASKING THAT ONE.
BUT NO. EVEN WEIRDER THAN THAT.

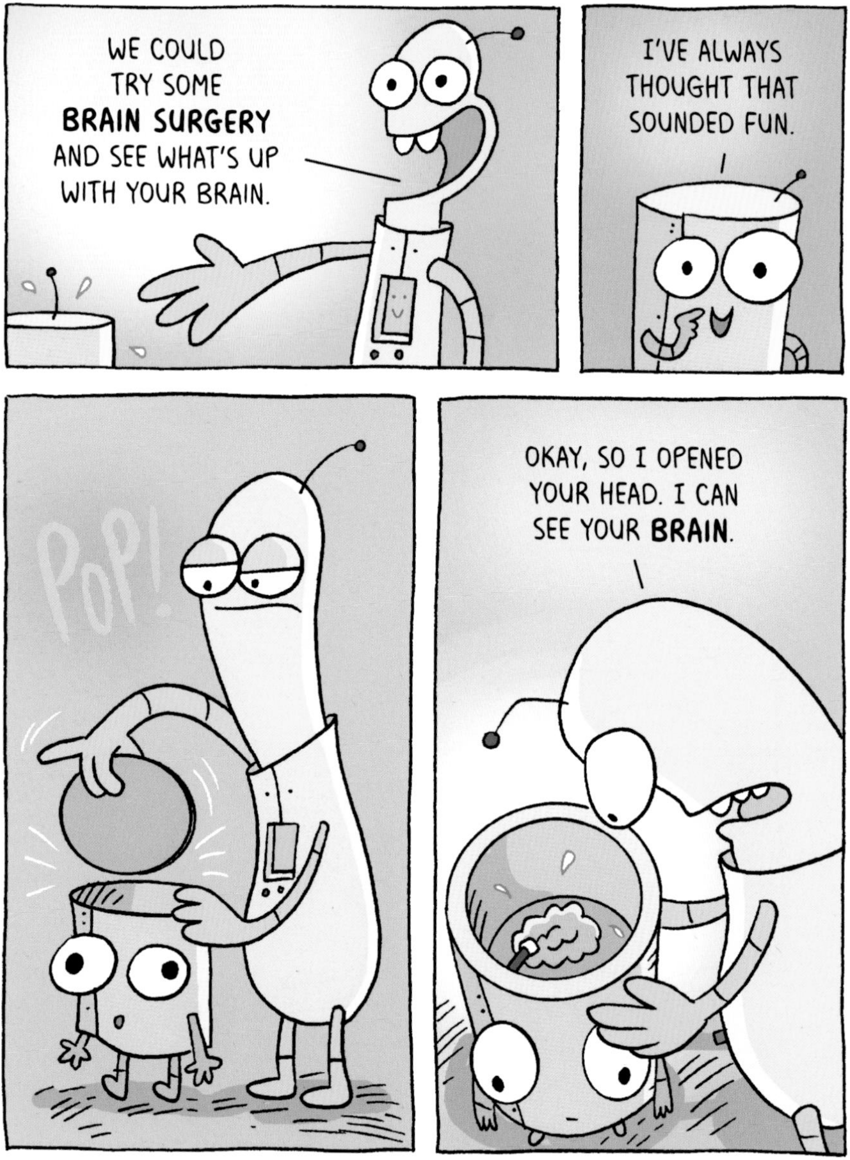
WE COULD TRY SOME **BRAIN SURGERY** AND SEE WHAT'S UP WITH YOUR BRAIN.
I'VE ALWAYS THOUGHT THAT SOUNDED FUN.
POP!
OKAY, SO I OPENED YOUR HEAD. I CAN SEE YOUR **BRAIN**.

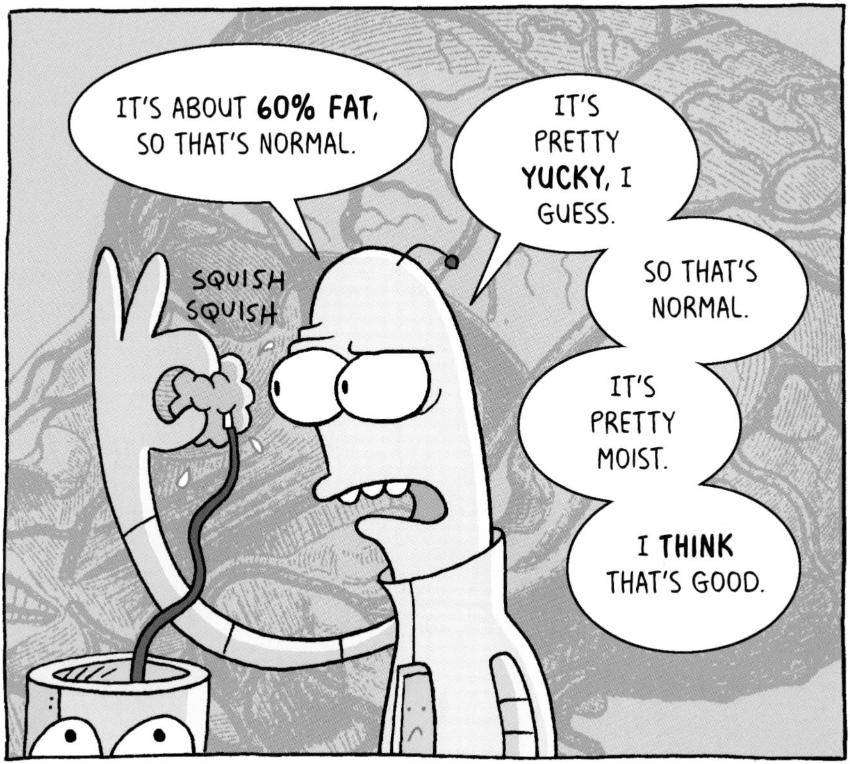
IT'S ABOUT **60% FAT,** SO THAT'S NORMAL.
IT'S PRETTY **YUCKY,** I GUESS.
SQUISH SQUISH
SO THAT'S NORMAL.
IT'S PRETTY MOIST.
I **THINK** THAT'S GOOD.

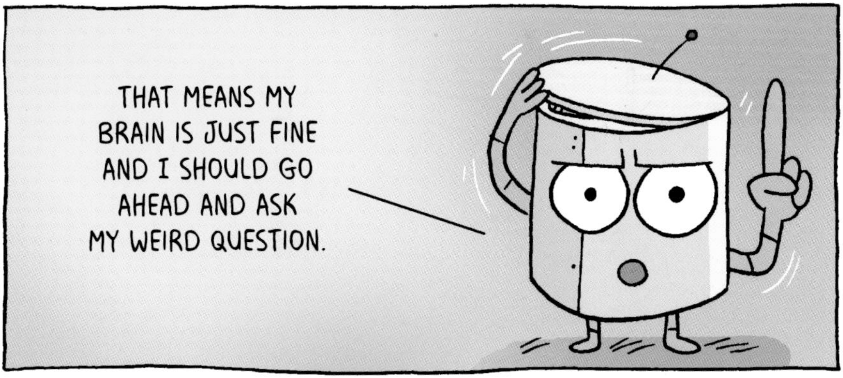
THAT MEANS MY BRAIN IS JUST FINE AND I SHOULD GO AHEAD AND ASK MY WEIRD QUESTION.

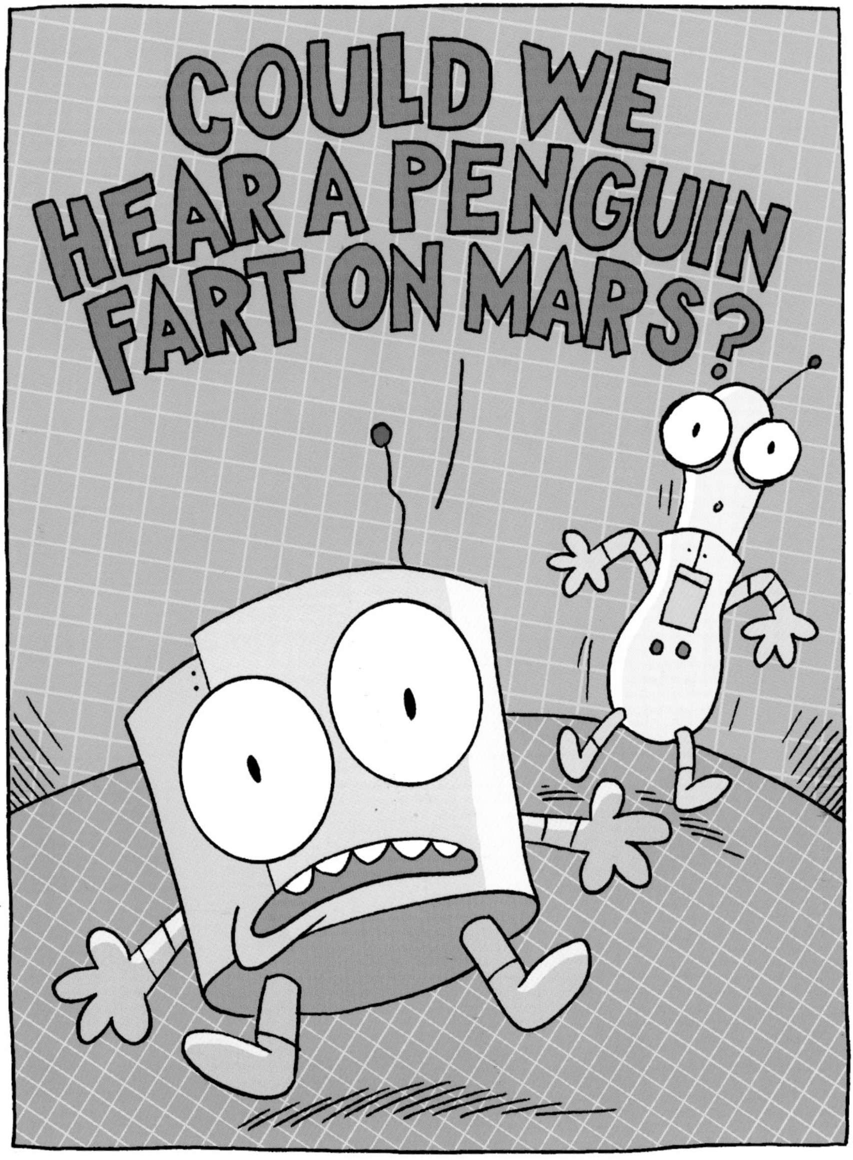
COULD WE HEAR A PENGUIN FART ON MARS?

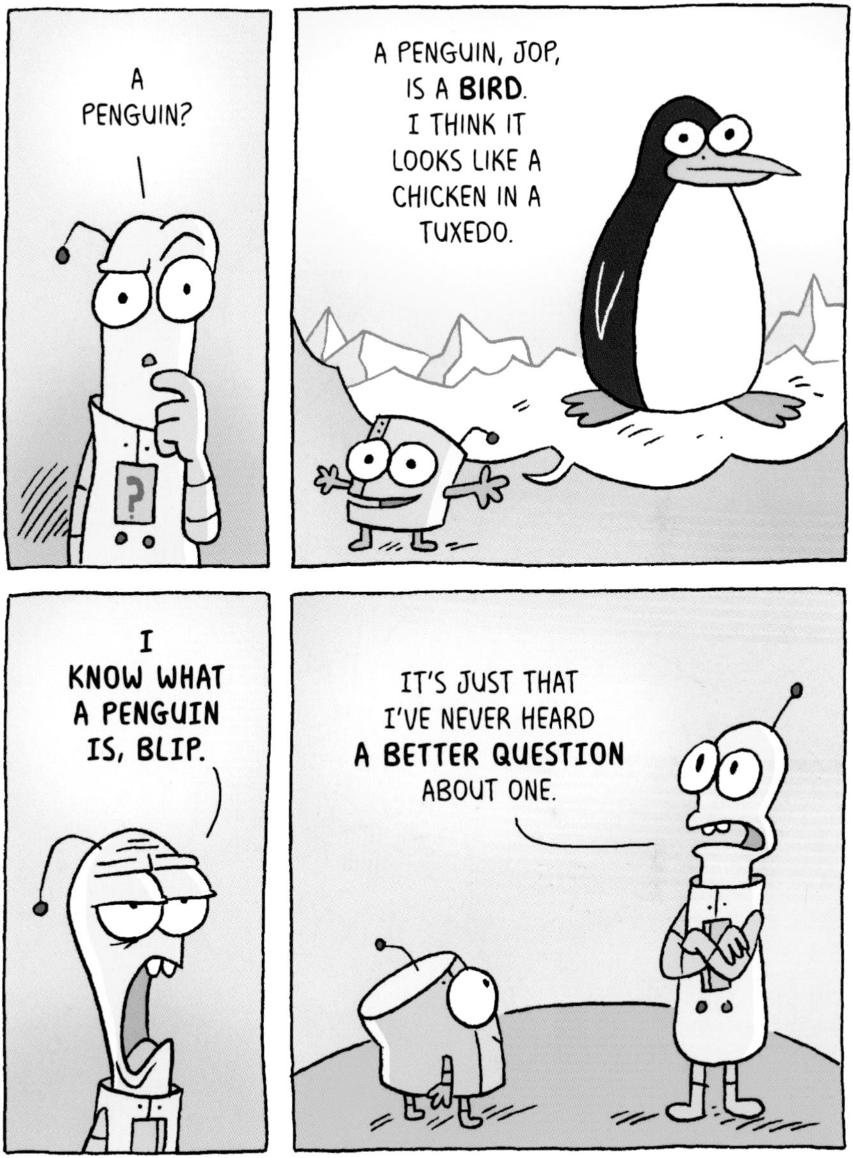
A PENGUIN?
A PENGUIN, JOP, IS A **BIRD**. I THINK IT LOOKS LIKE A CHICKEN IN A TUXEDO.
?
I KNOW WHAT A PENGUIN IS, BLIP.
IT'S JUST THAT I'VE NEVER HEARD **A BETTER QUESTION** ABOUT ONE.

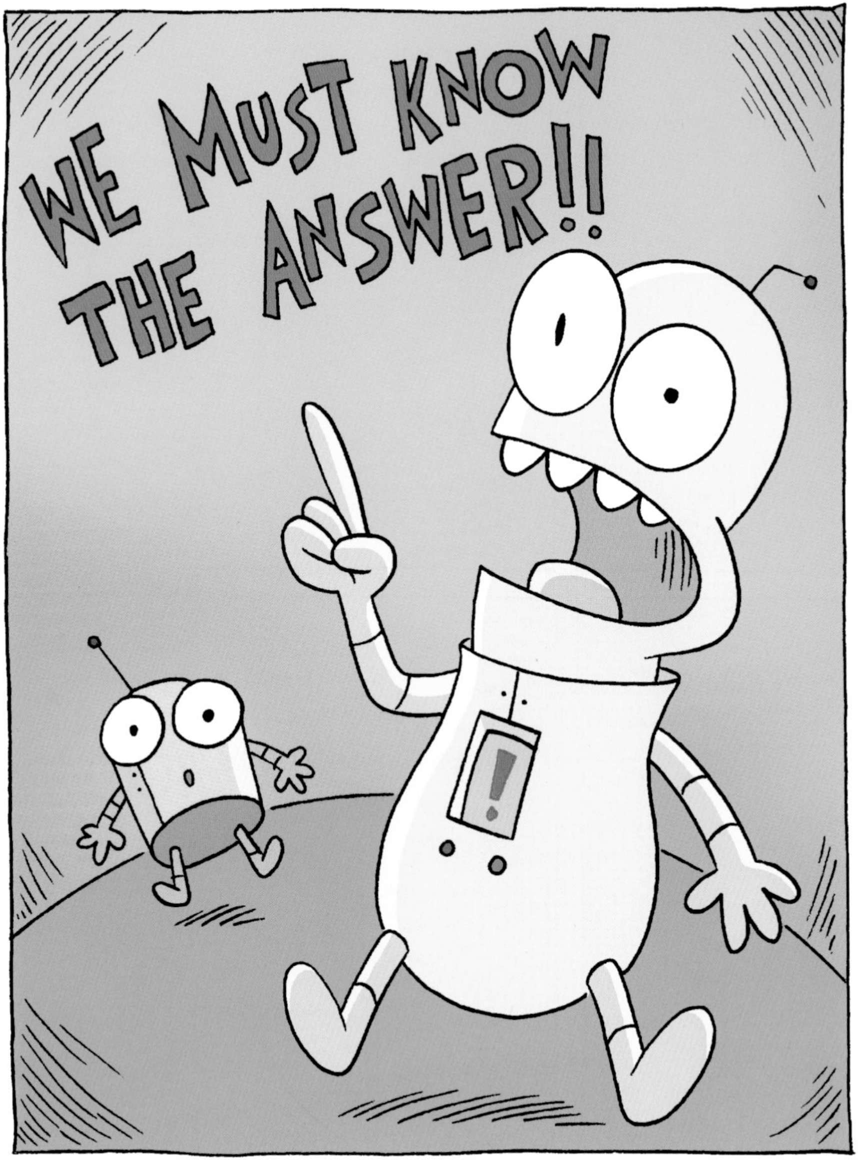
WE MUST KNOW THE ANSWER!!

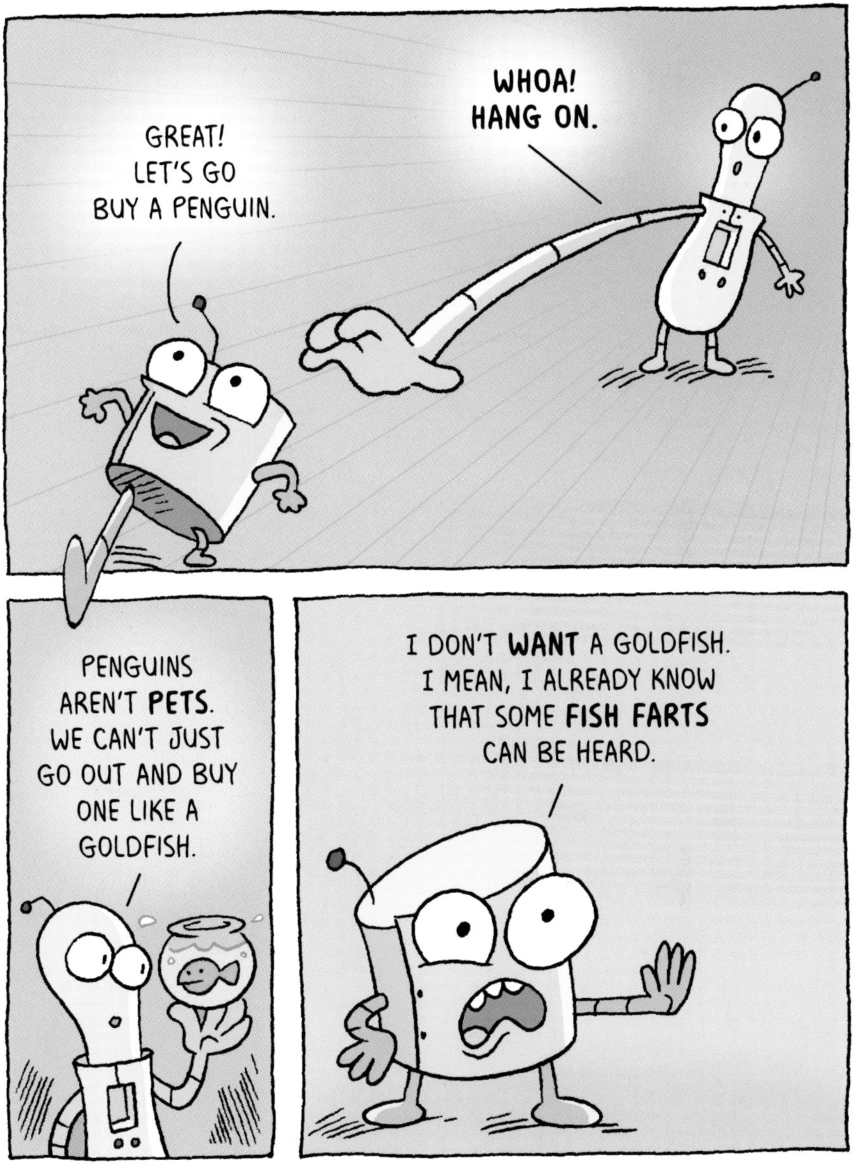
GREAT!
LET'S GO
BUY A PENGUIN.
WHOA!
HANG ON.
PENGUINS
AREN'T PETS.
WE CAN'T JUST
GO OUT AND BUY
ONE LIKE A
GOLDFISH.
I DON'T WANT A GOLDFISH.
I MEAN, I ALREADY KNOW
THAT SOME FISH FARTS
CAN BE HEARD.

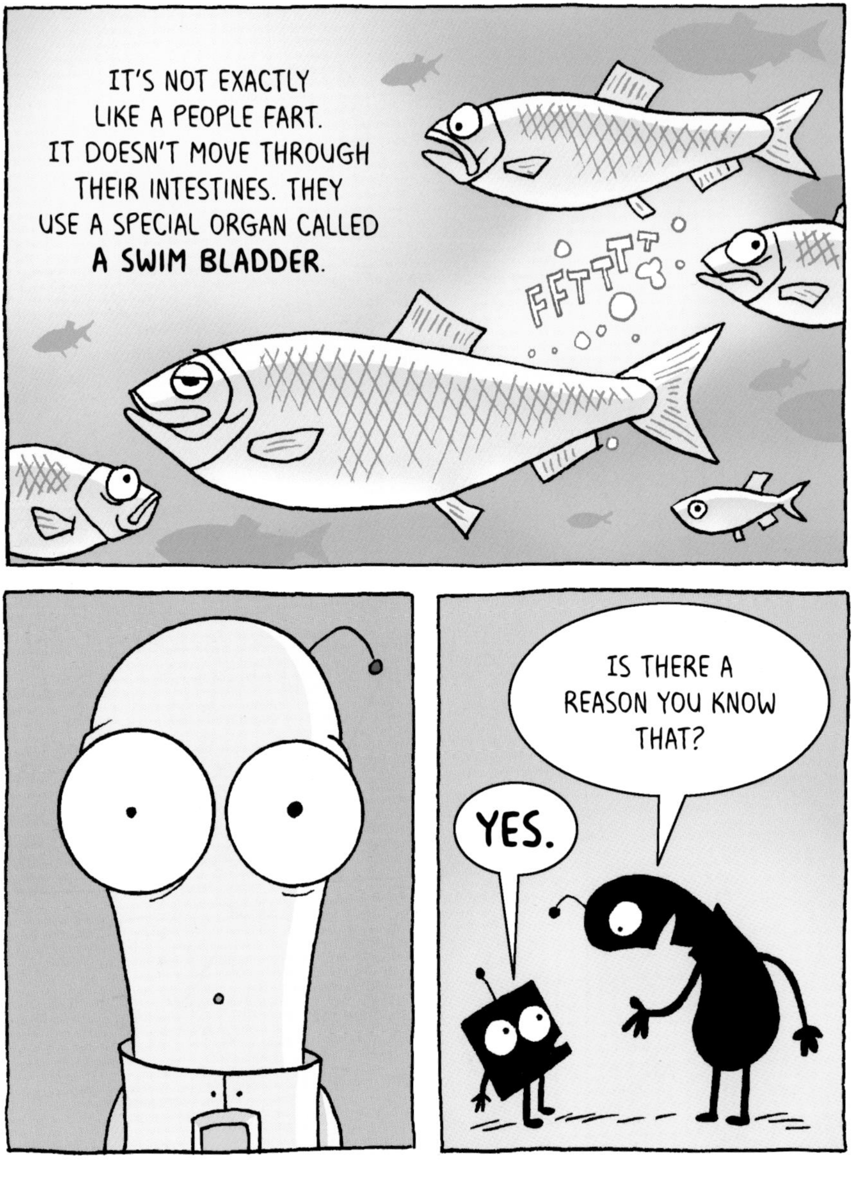
IT'S NOT EXACTLY LIKE A PEOPLE FART. IT DOESN'T MOVE THROUGH THEIR INTESTINES. THEY USE A SPECIAL ORGAN CALLED **A SWIM BLADDER**.
FFTTTTT
IS THERE A REASON YOU KNOW THAT?
YES.

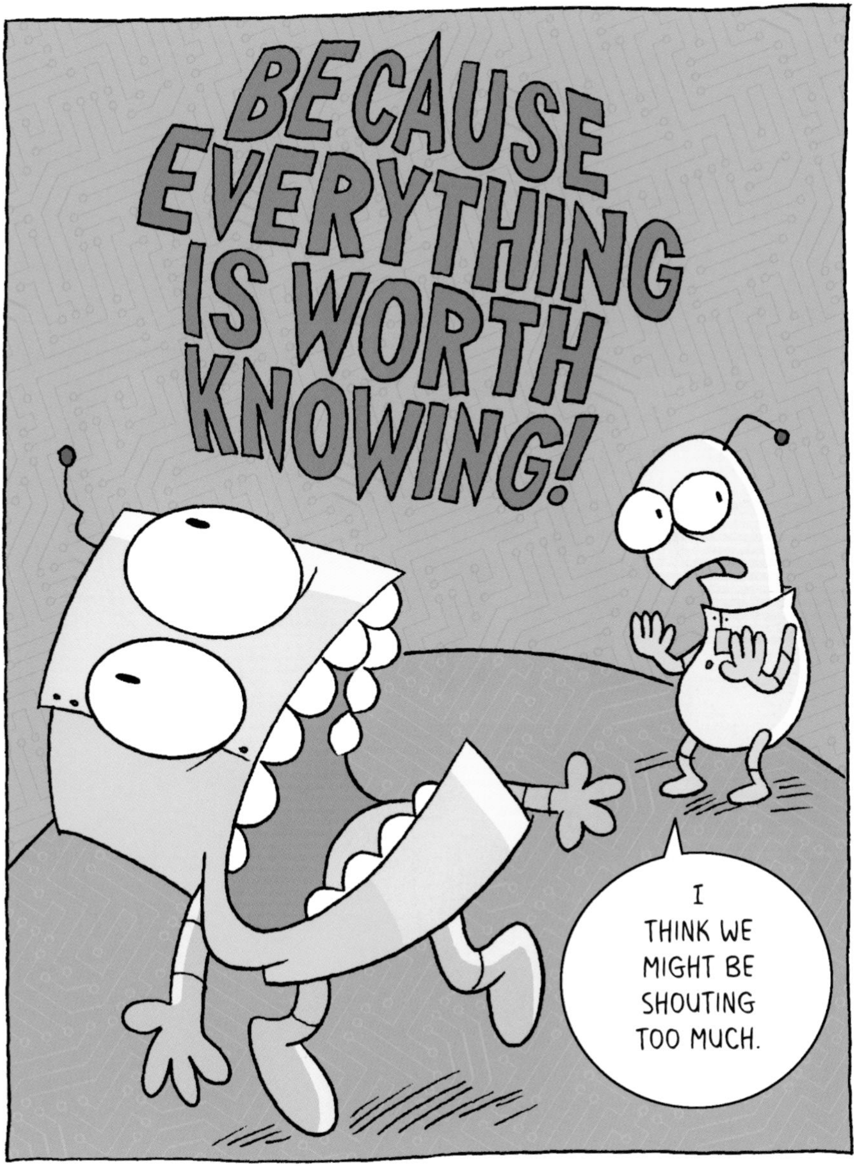
BECAUSE EVERYTHING IS WORTH KNOWING!
I THINK WE MIGHT BE SHOUTING TOO MUCH.

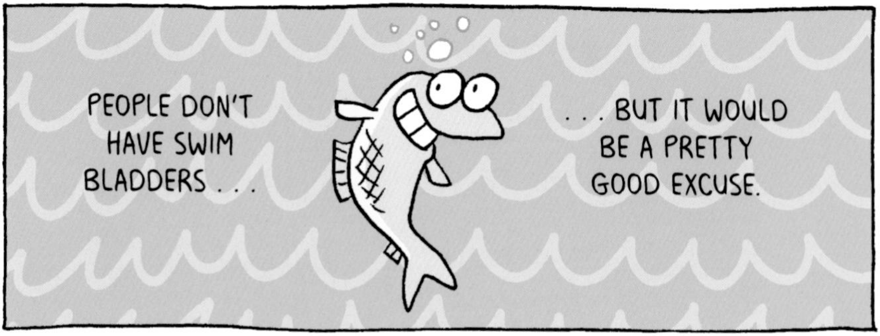
PEOPLE DON'T HAVE SWIM BLADDERS . . .
. . . BUT IT WOULD BE A PRETTY GOOD EXCUSE.

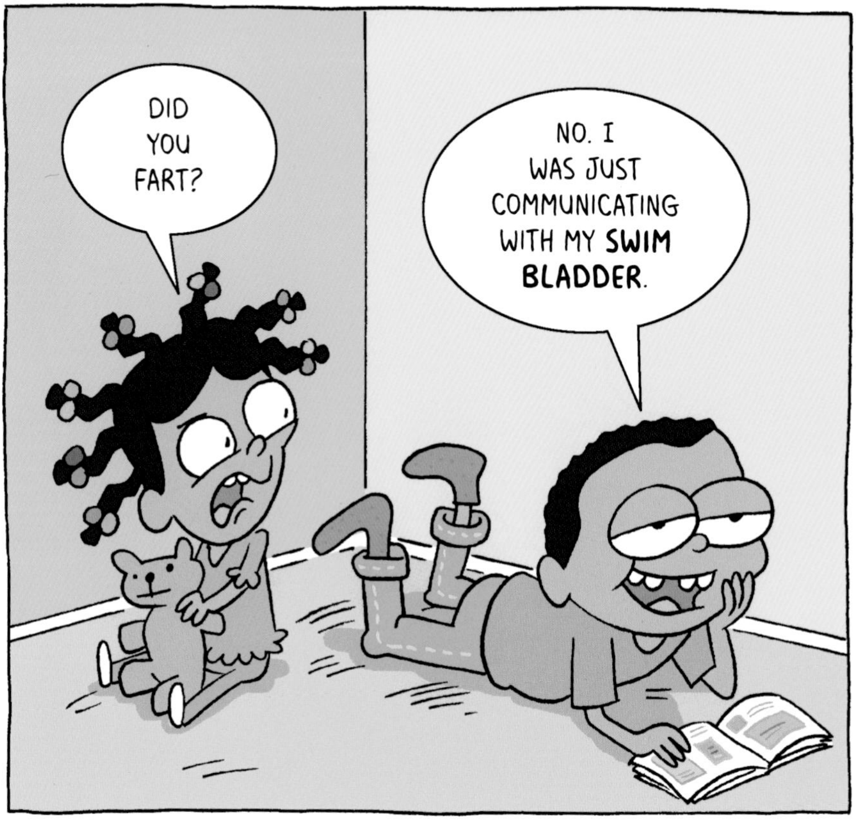
DID YOU FART?
NO. I WAS JUST COMMUNICATING WITH MY **SWIM BLADDER**.

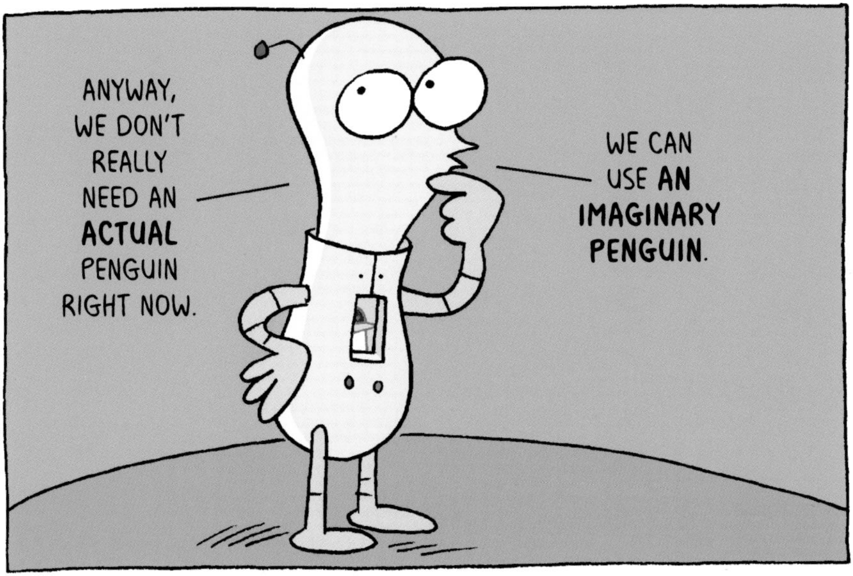
ANYWAY, WE DON'T REALLY NEED AN **ACTUAL** PENGUIN RIGHT NOW.
WE CAN USE **AN IMAGINARY PENGUIN**.

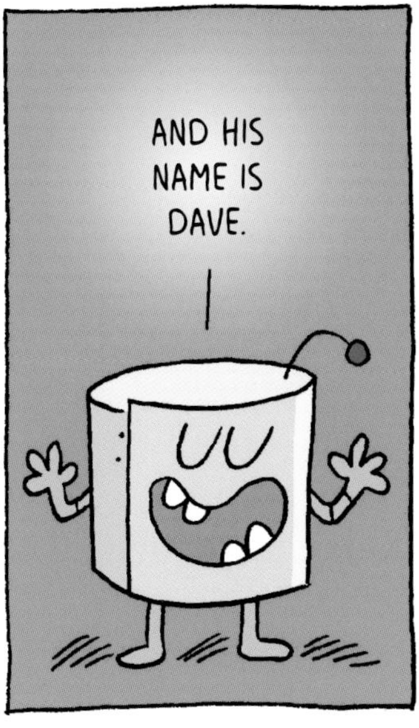
AND HIS NAME IS DAVE.

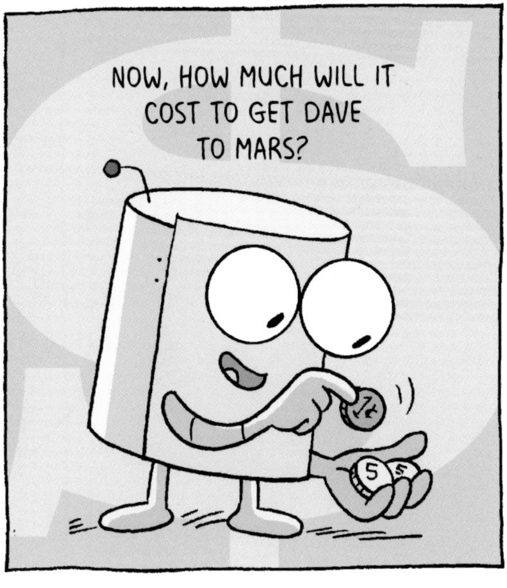
NOW, HOW MUCH WILL IT COST TO GET DAVE TO MARS?

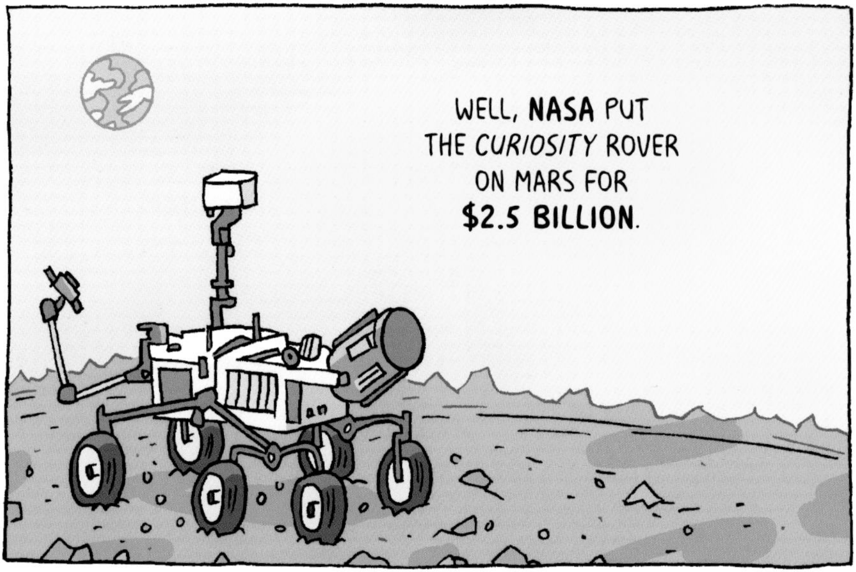
WELL, **NASA** PUT THE *CURIOSITY* ROVER ON MARS FOR **$2.5 BILLION**.

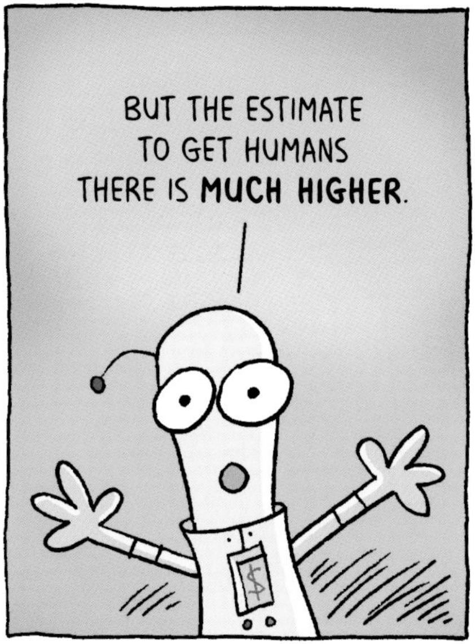
BUT THE ESTIMATE TO GET HUMANS THERE IS **MUCH HIGHER**.

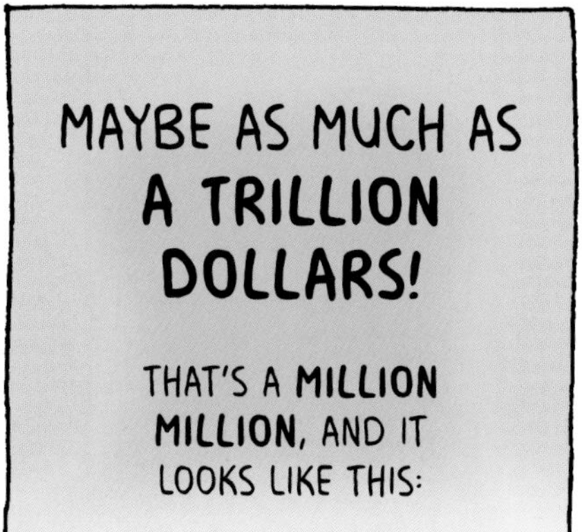
MAYBE AS MUCH AS **A TRILLION DOLLARS!**
THAT'S A **MILLION MILLION**, AND IT LOOKS LIKE THIS:
$1,000,000,000,000.00

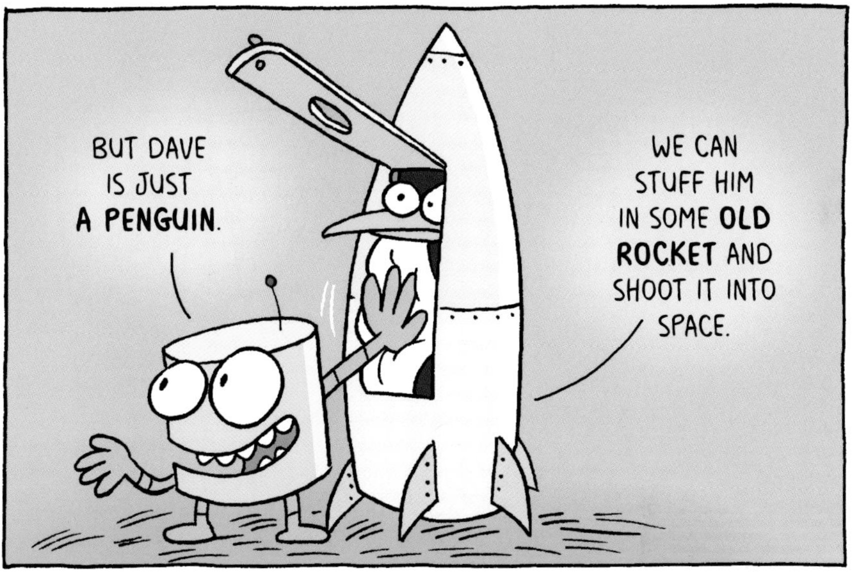
BUT DAVE IS JUST **A PENGUIN**.
WE CAN STUFF HIM IN SOME **OLD ROCKET** AND SHOOT IT INTO SPACE.

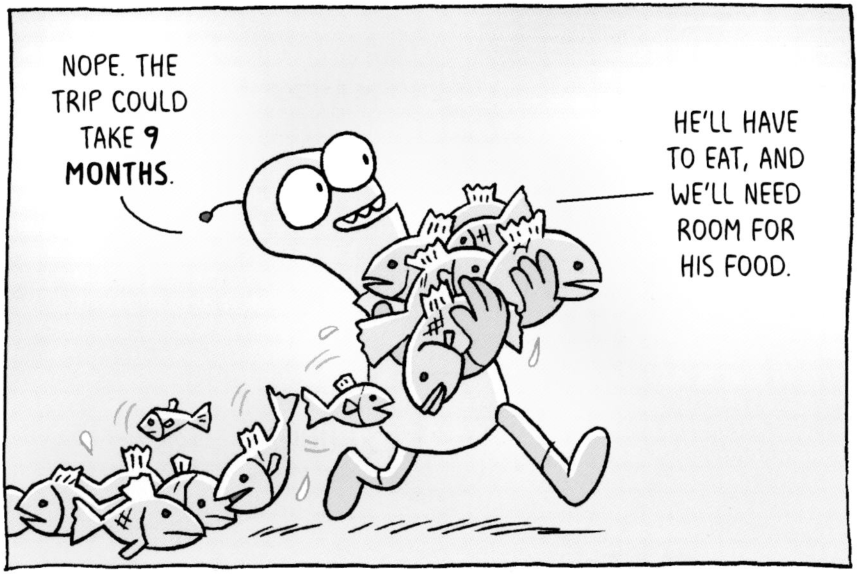
NOPE. THE TRIP COULD TAKE **9 MONTHS**.
HE'LL HAVE TO EAT, AND WE'LL NEED ROOM FOR HIS FOOD.

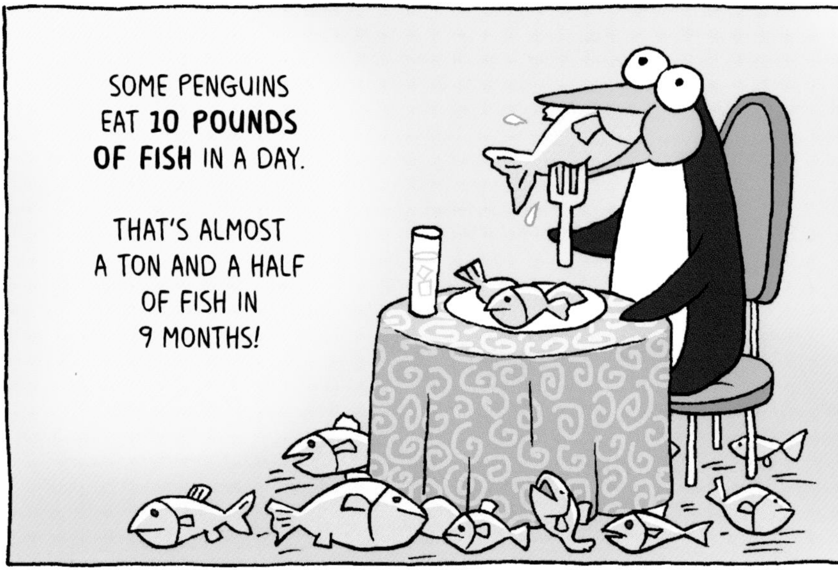
SOME PENGUINS EAT **10 POUNDS OF FISH** IN A DAY.
THAT'S ALMOST A TON AND A HALF OF FISH IN 9 MONTHS!

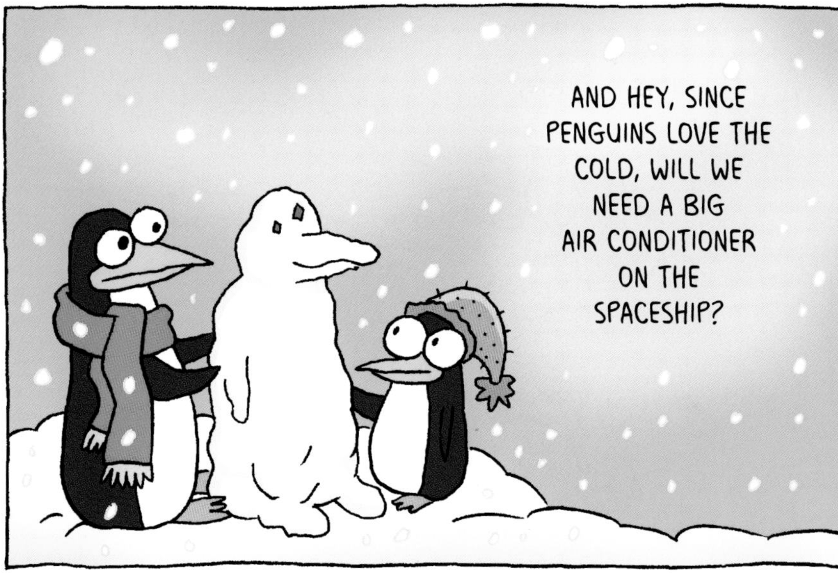
AND HEY, SINCE PENGUINS LOVE THE COLD, WILL WE NEED A BIG AIR CONDITIONER ON THE SPACESHIP?

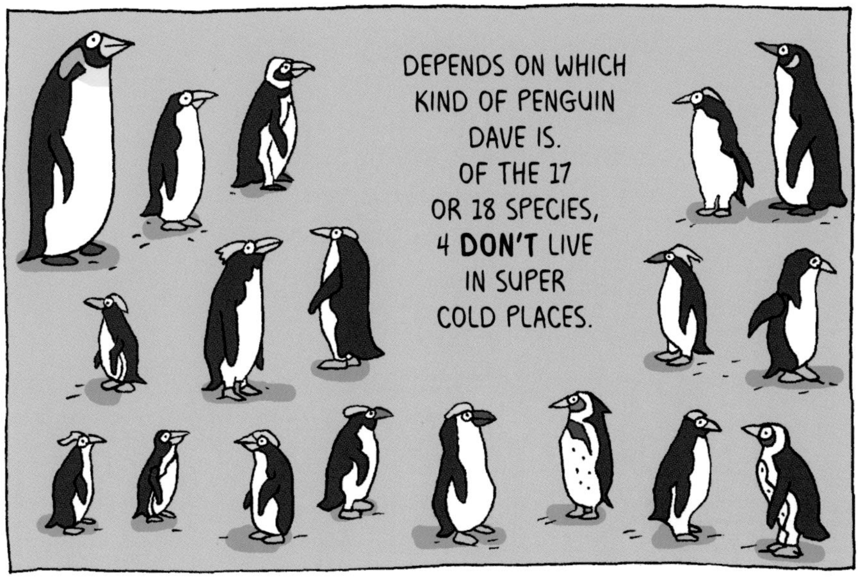
DEPENDS ON WHICH KIND OF PENGUIN DAVE IS. OF THE 17 OR 18 SPECIES, 4 **DON'T** LIVE IN SUPER COLD PLACES.

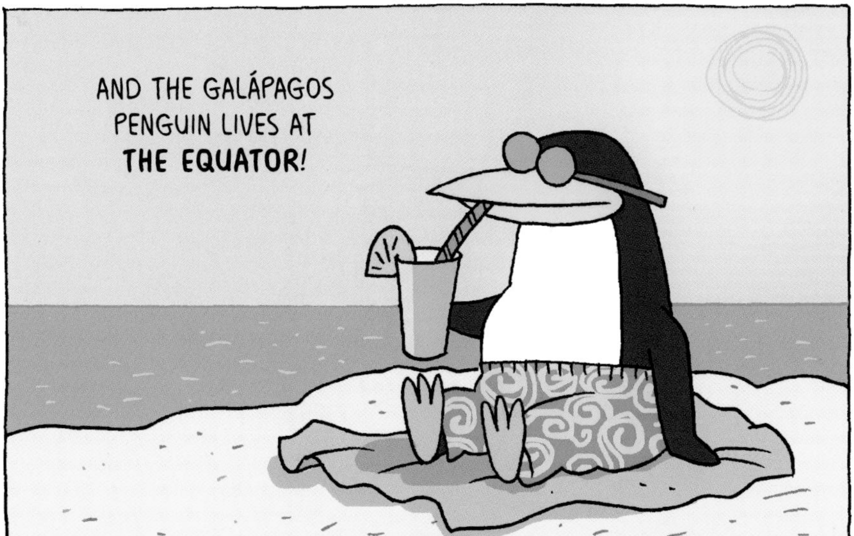
AND THE GALÁPAGOS PENGUIN LIVES AT **THE EQUATOR**!

CAN YOU FIND THE TWO JOPS THAT ARE **EXACTLY THE SAME?**

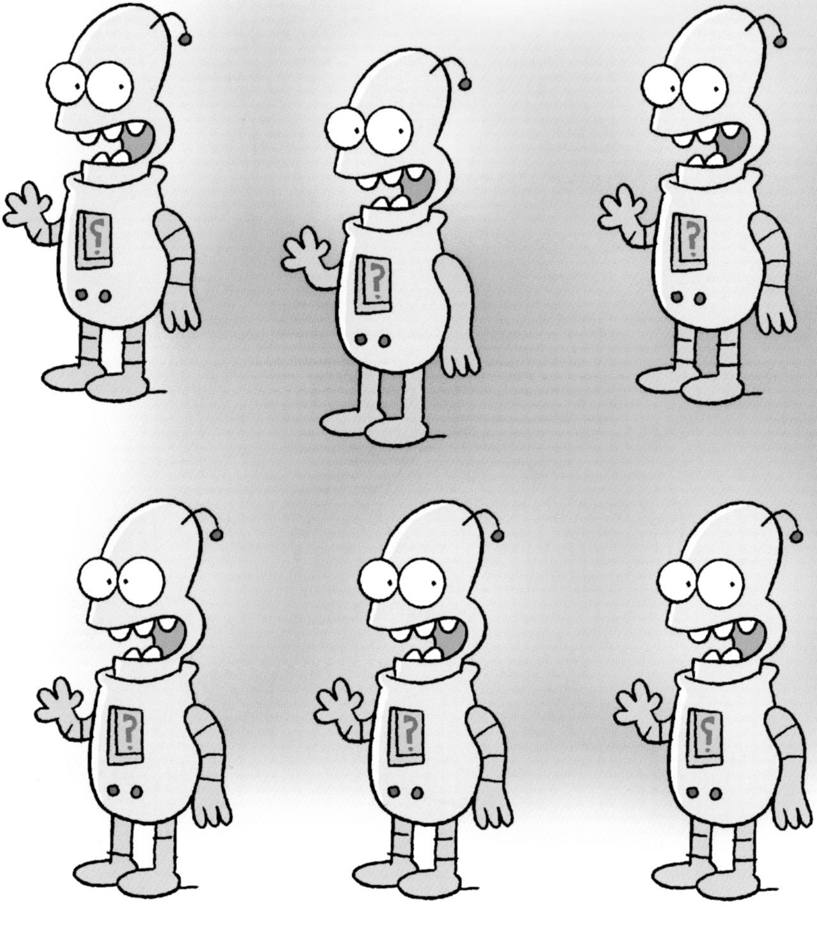

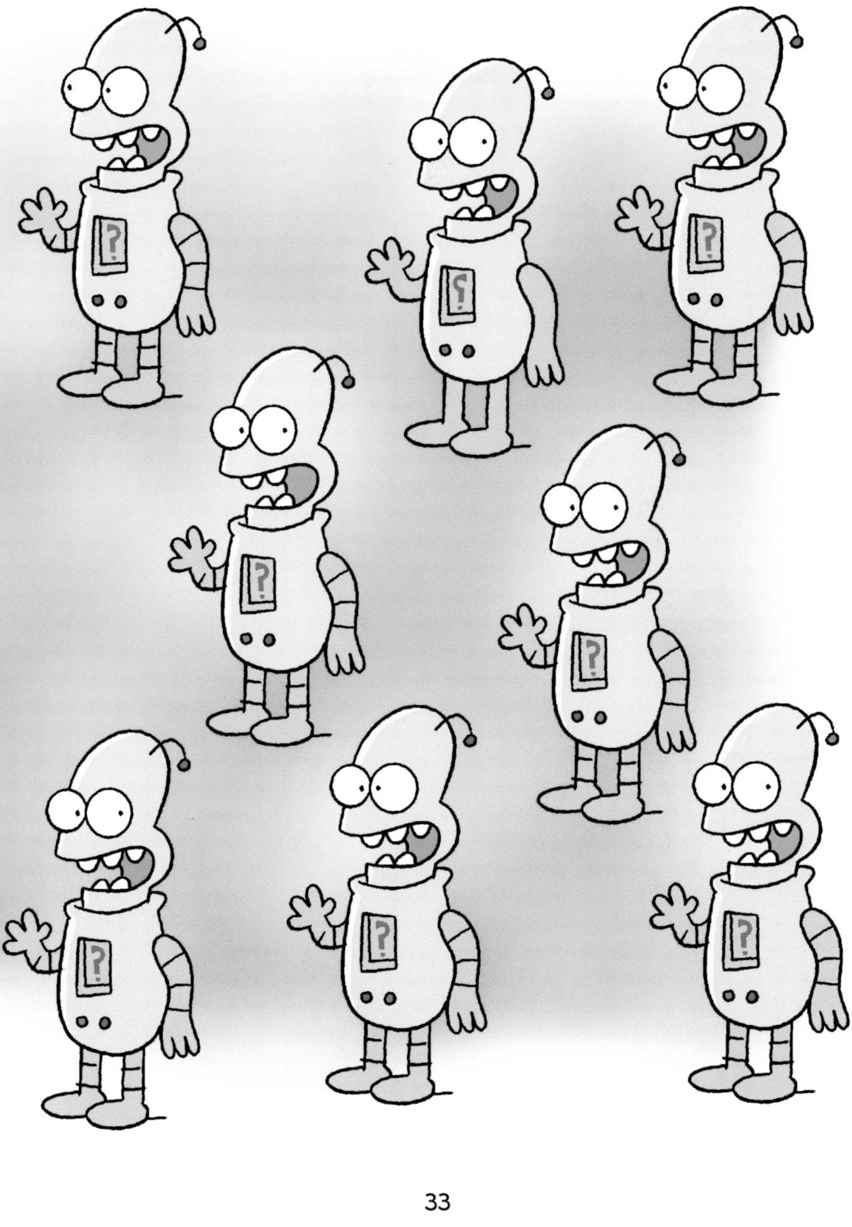

THE DRAGON DILEMMA

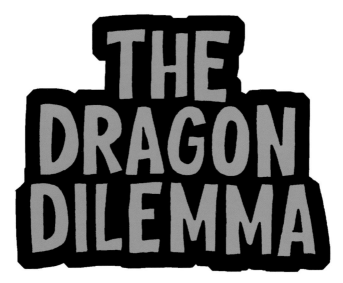

43

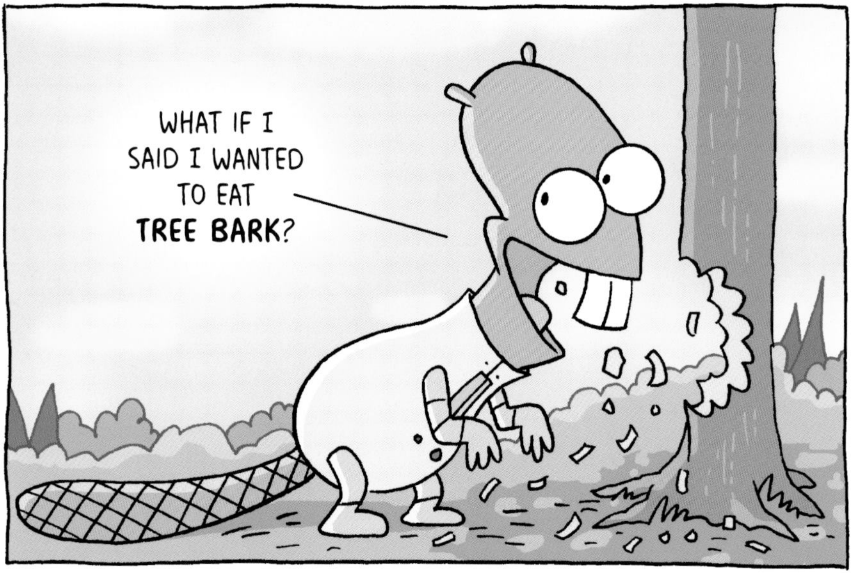
WHAT IF I SAID I WANTED TO EAT **TREE BARK**?

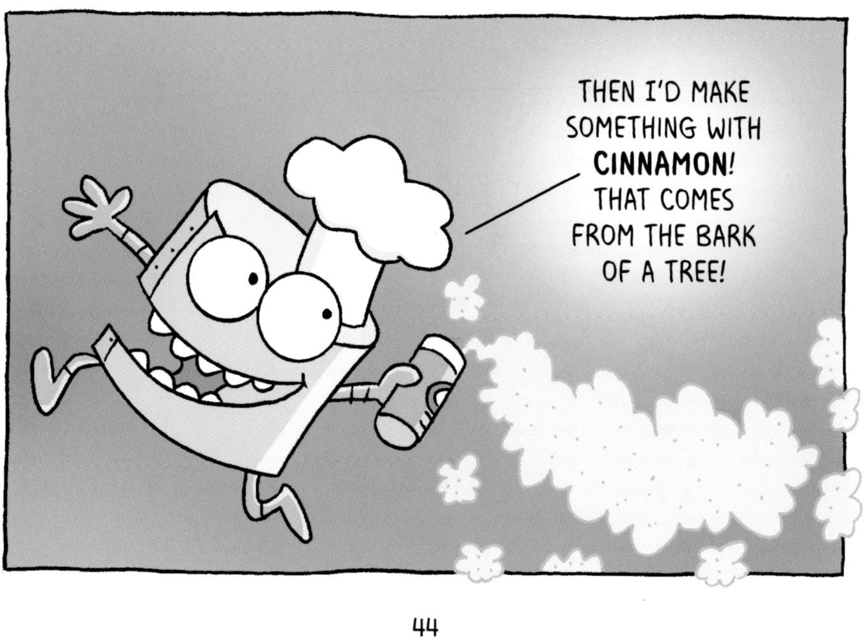
THEN I'D MAKE SOMETHING WITH **CINNAMON**! THAT COMES FROM THE BARK OF A TREE!

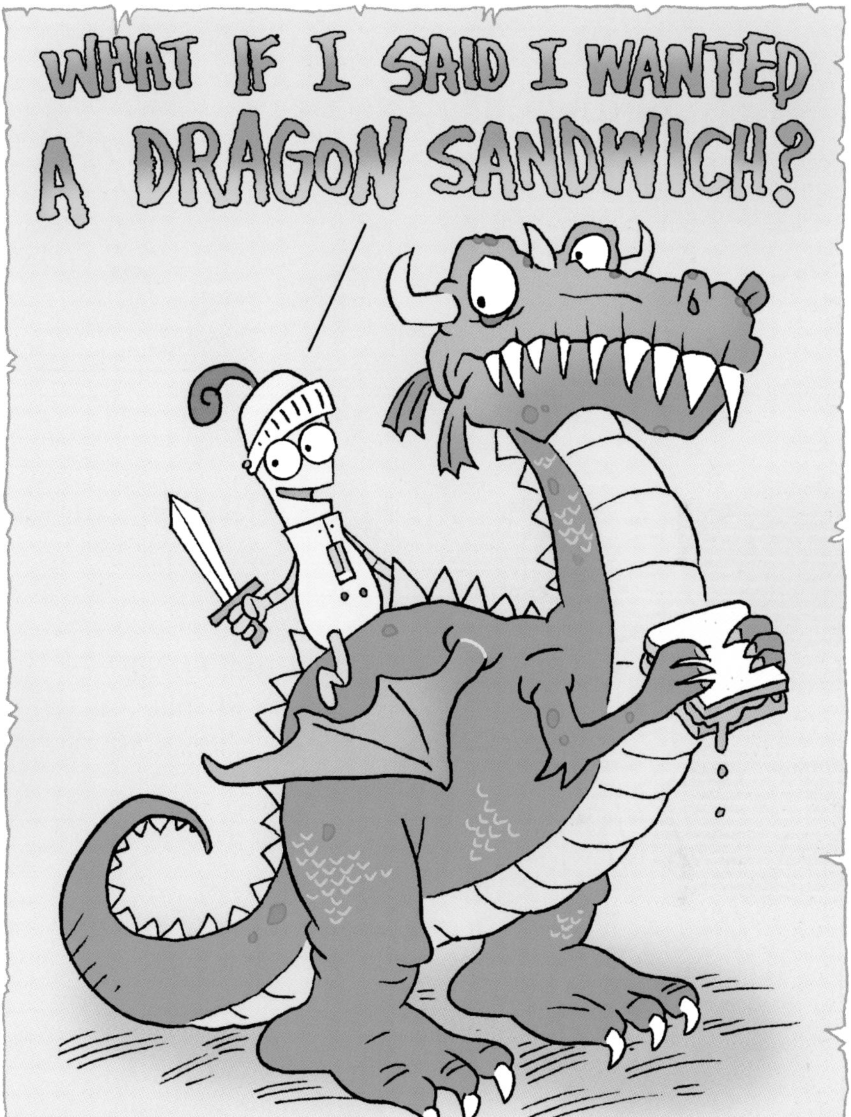
WHAT IF I SAID I WANTED A DRAGON SANDWICH?

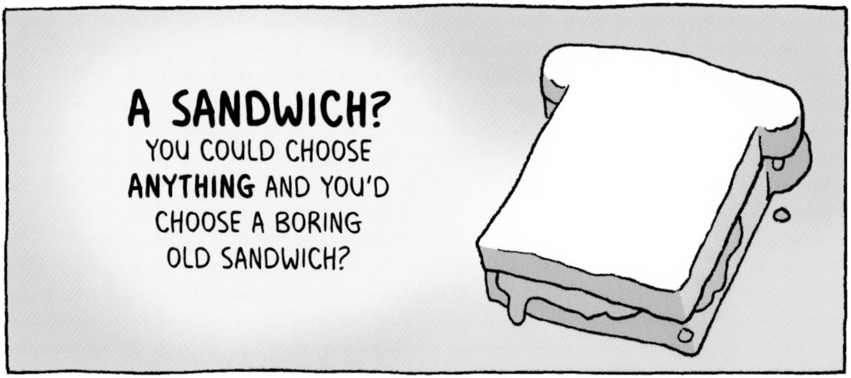
A SANDWICH?
YOU COULD CHOOSE ANYTHING AND YOU'D CHOOSE A BORING OLD SANDWICH?

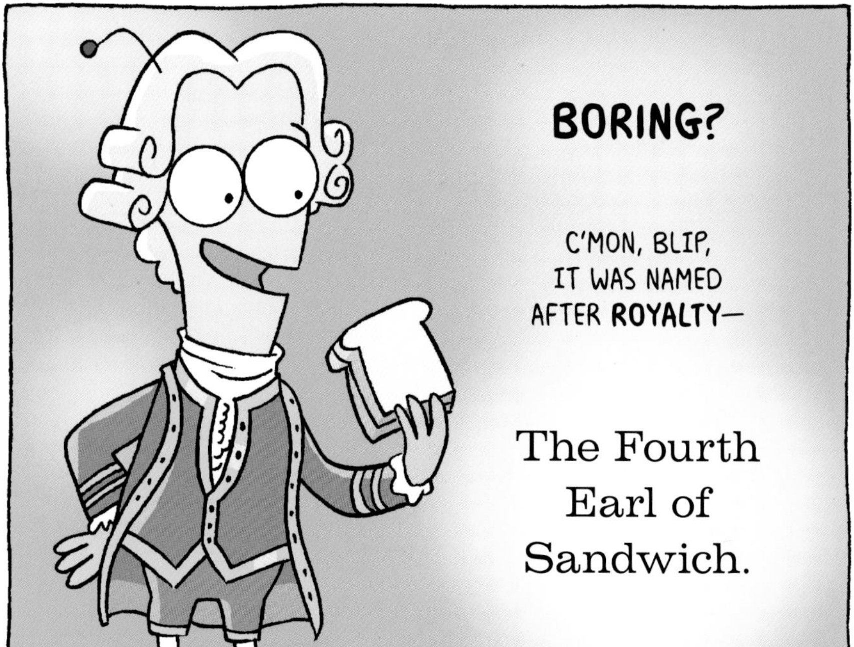
BORING?
C'MON, BLIP, IT WAS NAMED AFTER ROYALTY—
The Fourth Earl of Sandwich.
(SANDWICH IS A TOWN IN ENGLAND.)

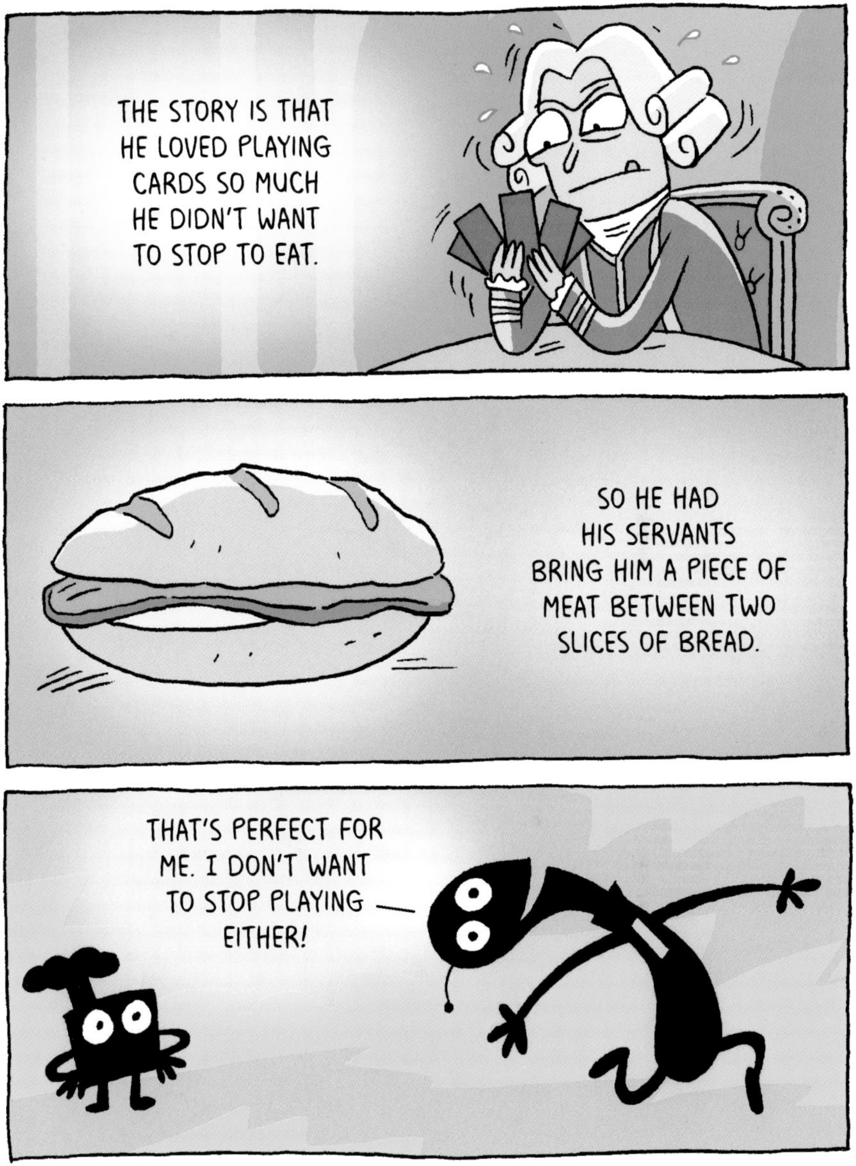
THE STORY IS THAT HE LOVED PLAYING CARDS SO MUCH HE DIDN'T WANT TO STOP TO EAT.
SO HE HAD HIS SERVANTS BRING HIM A PIECE OF MEAT BETWEEN TWO SLICES OF BREAD.
THAT'S PERFECT FOR ME. I DON'T WANT TO STOP PLAYING EITHER!

48

SOME PEOPLE SAY THAT THE FIRST HOT DOG SELLER LOANED YOU **A GLOVE** TO WEAR WHILE YOU ATE IT.

BUT PEOPLE KEPT RUNNING OFF WITH THE GLOVES, SO THEY CAME UP WITH THE IDEA OF A **BUN**.

53

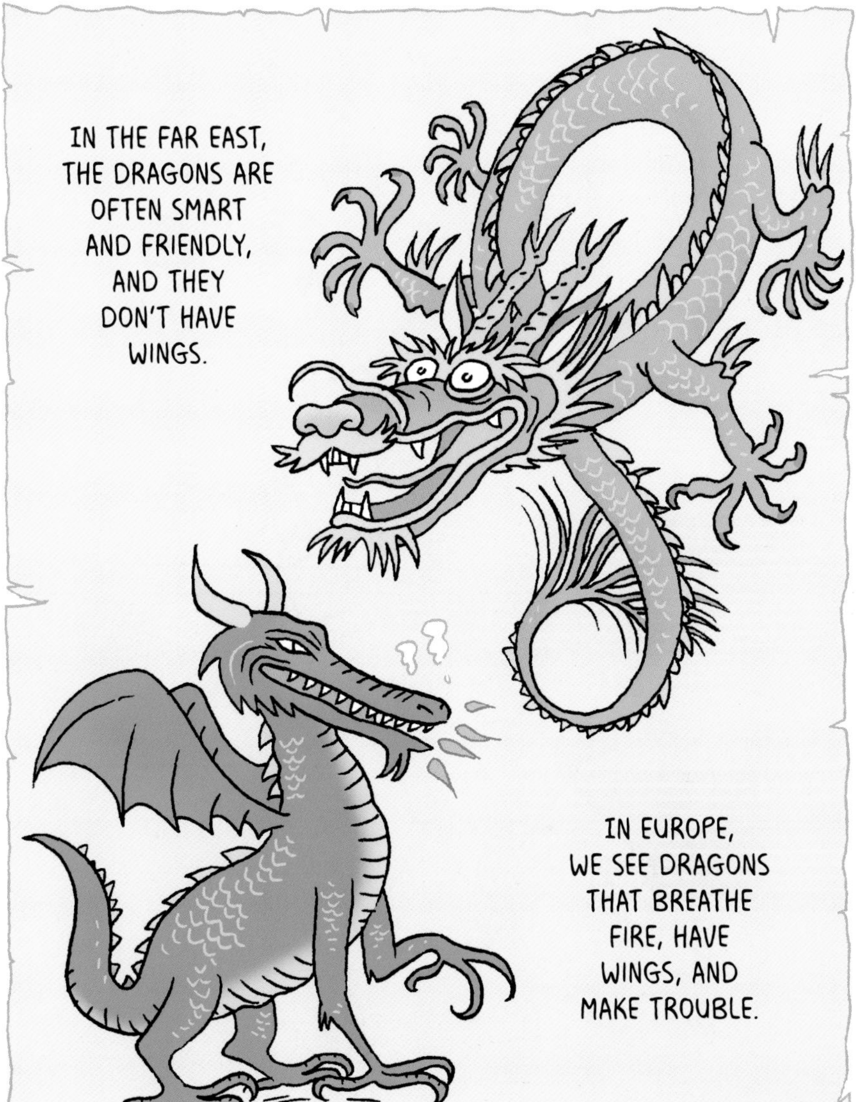
IN THE FAR EAST, THE DRAGONS ARE OFTEN SMART AND FRIENDLY, AND THEY DON'T HAVE WINGS.
IN EUROPE, WE SEE DRAGONS THAT BREATHE FIRE, HAVE WINGS, AND MAKE TROUBLE.

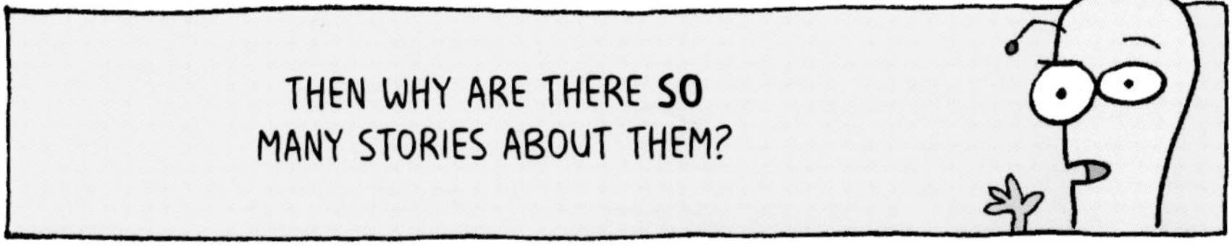
THEN WHY ARE THERE **SO** MANY STORIES ABOUT THEM?

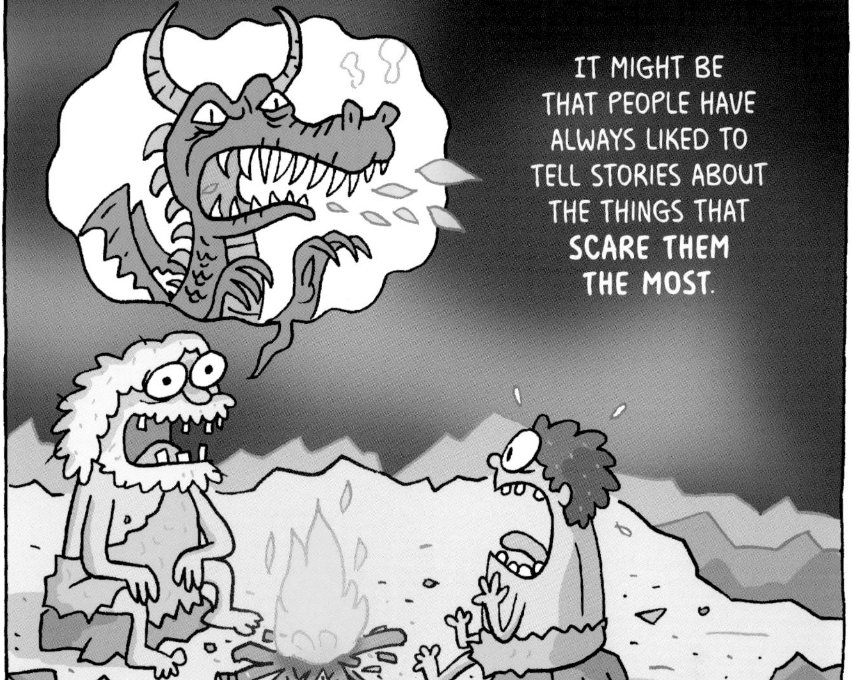
IT MIGHT BE THAT PEOPLE HAVE ALWAYS LIKED TO TELL STORIES ABOUT THE THINGS THAT **SCARE THEM THE MOST**.

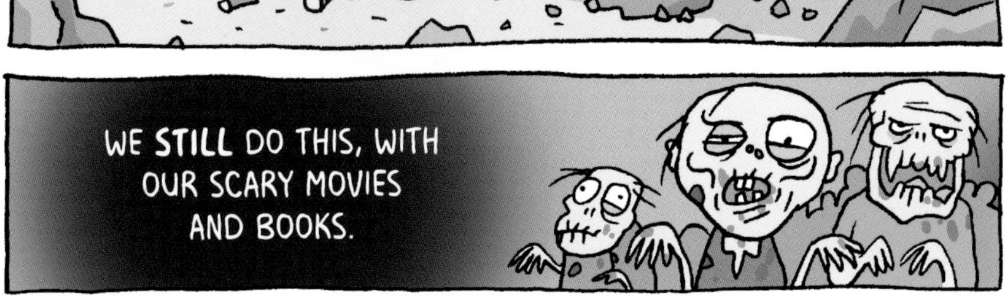
WE **STILL** DO THIS, WITH OUR SCARY MOVIES AND BOOKS.

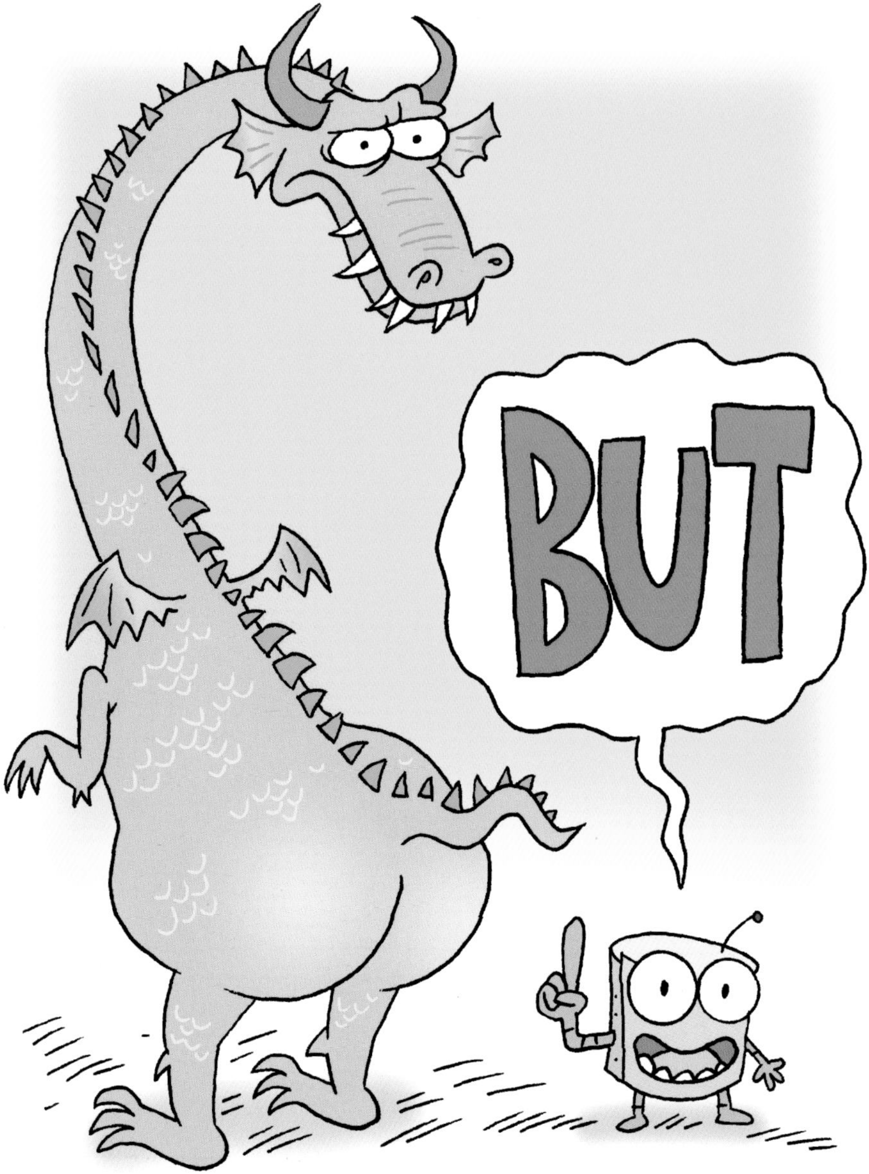
BUT

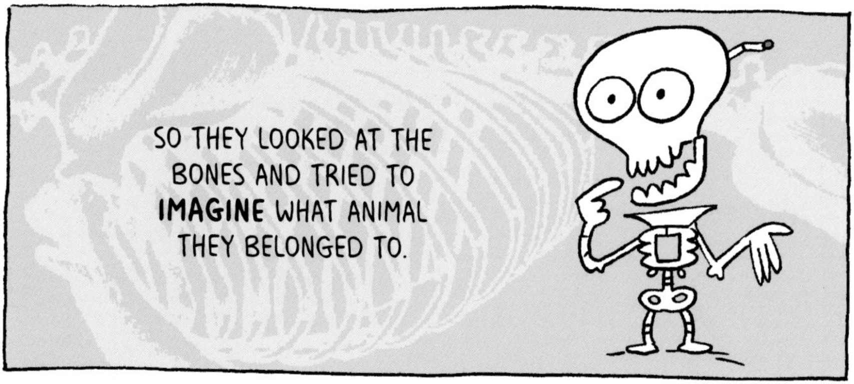
SO THEY LOOKED AT THE BONES AND TRIED TO **IMAGINE** WHAT ANIMAL THEY BELONGED TO.

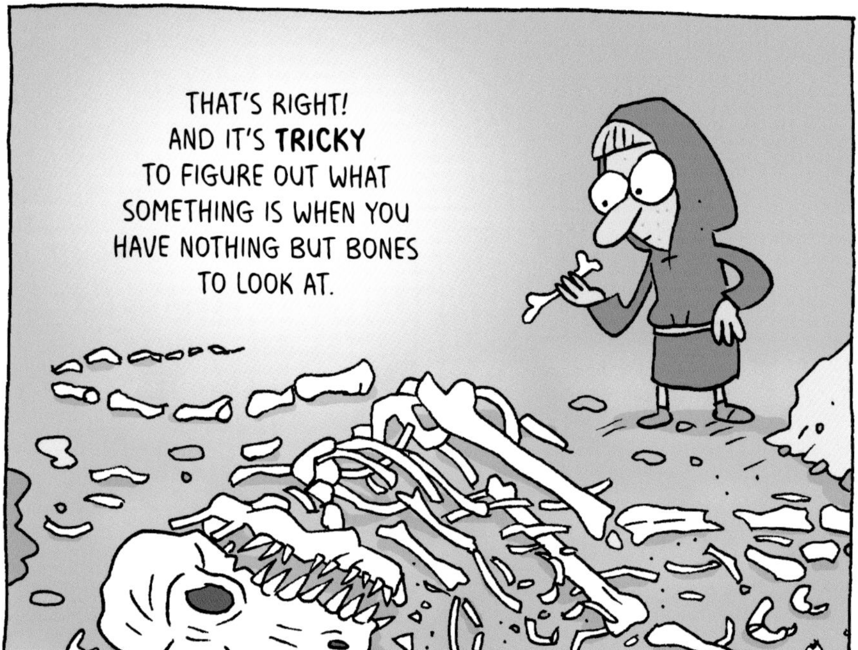
THAT'S RIGHT! AND IT'S **TRICKY** TO FIGURE OUT WHAT SOMETHING IS WHEN YOU HAVE NOTHING BUT BONES TO LOOK AT.

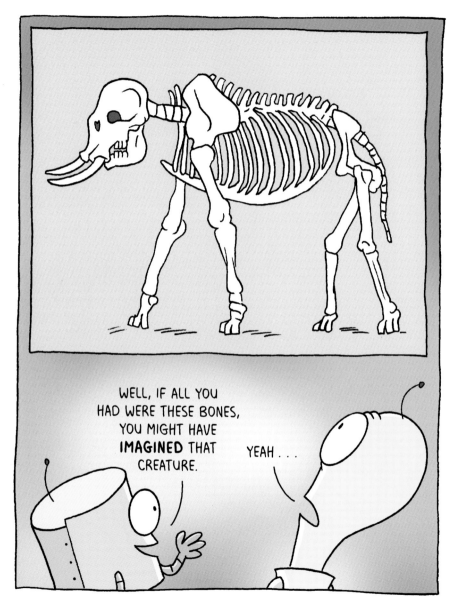

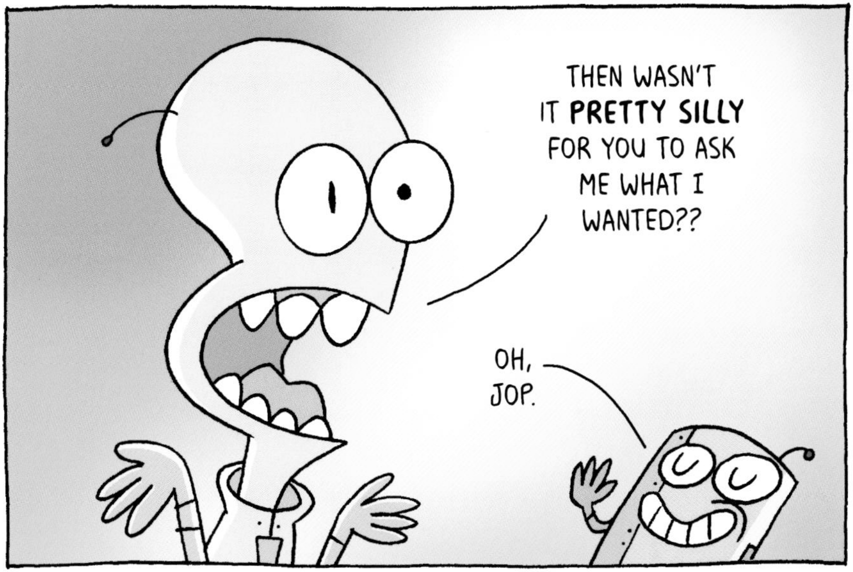
THEN WASN'T IT **PRETTY SILLY** FOR YOU TO ASK ME WHAT I WANTED??
OH, JOP.

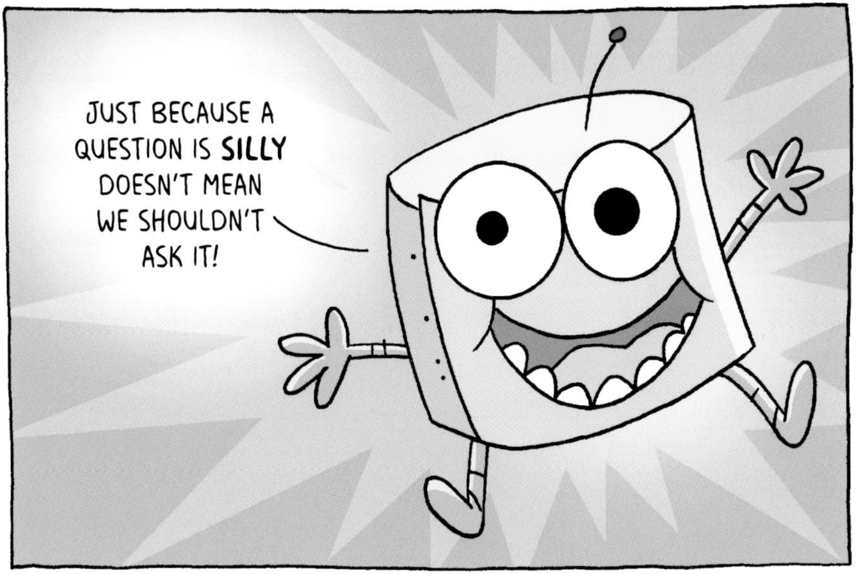
JUST BECAUSE A QUESTION IS **SILLY** DOESN'T MEAN WE SHOULDN'T ASK IT!

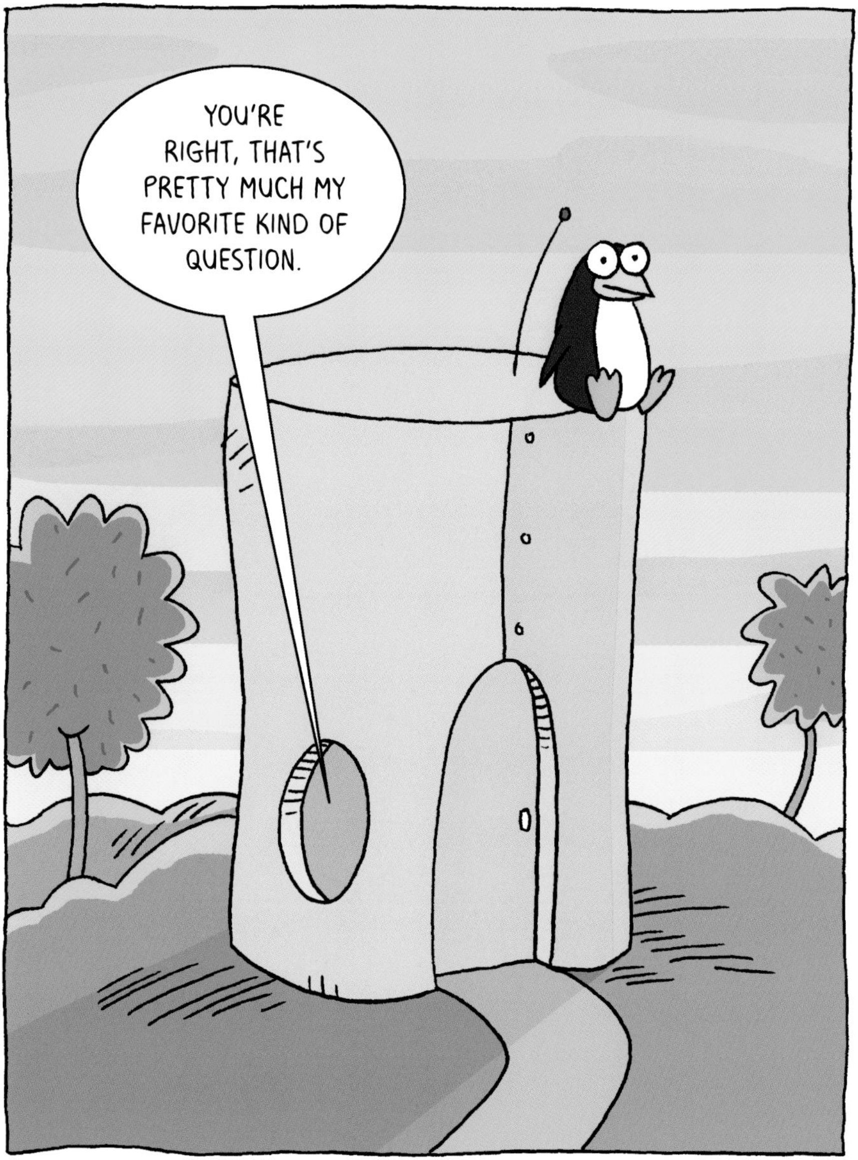
YOU'RE RIGHT, THAT'S PRETTY MUCH MY FAVORITE KIND OF QUESTION.

YOU KNOW THAT A DINOSAUR COULD BE ANY
COLOR AND MIGHT HAVE HAD FEATHERS.

TRACE AND COLOR THIS ONE ANY WAY **YOU** WANT TO.

THE SENSORY SURPRISE

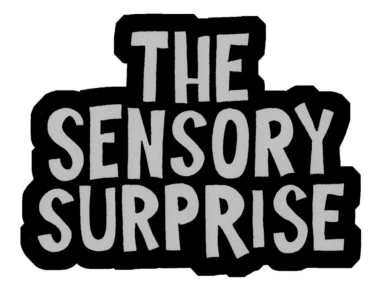

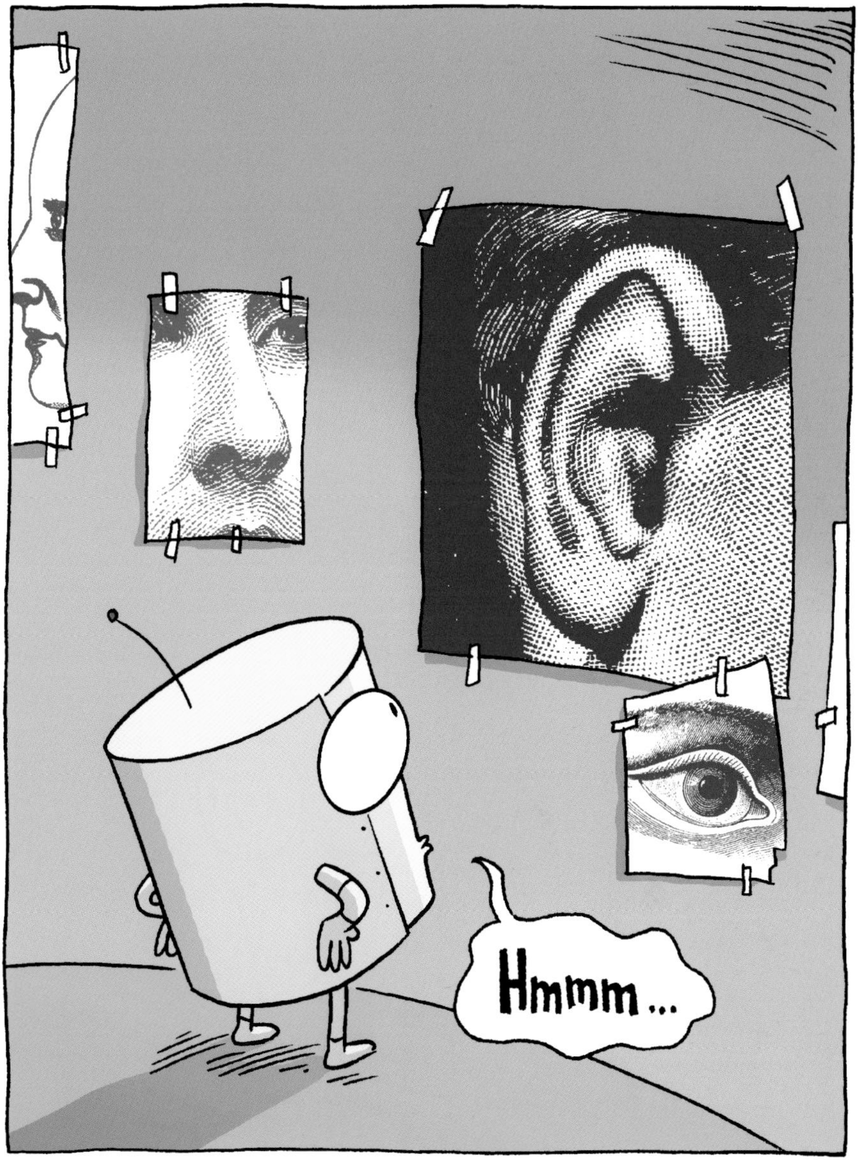
Hmmm...

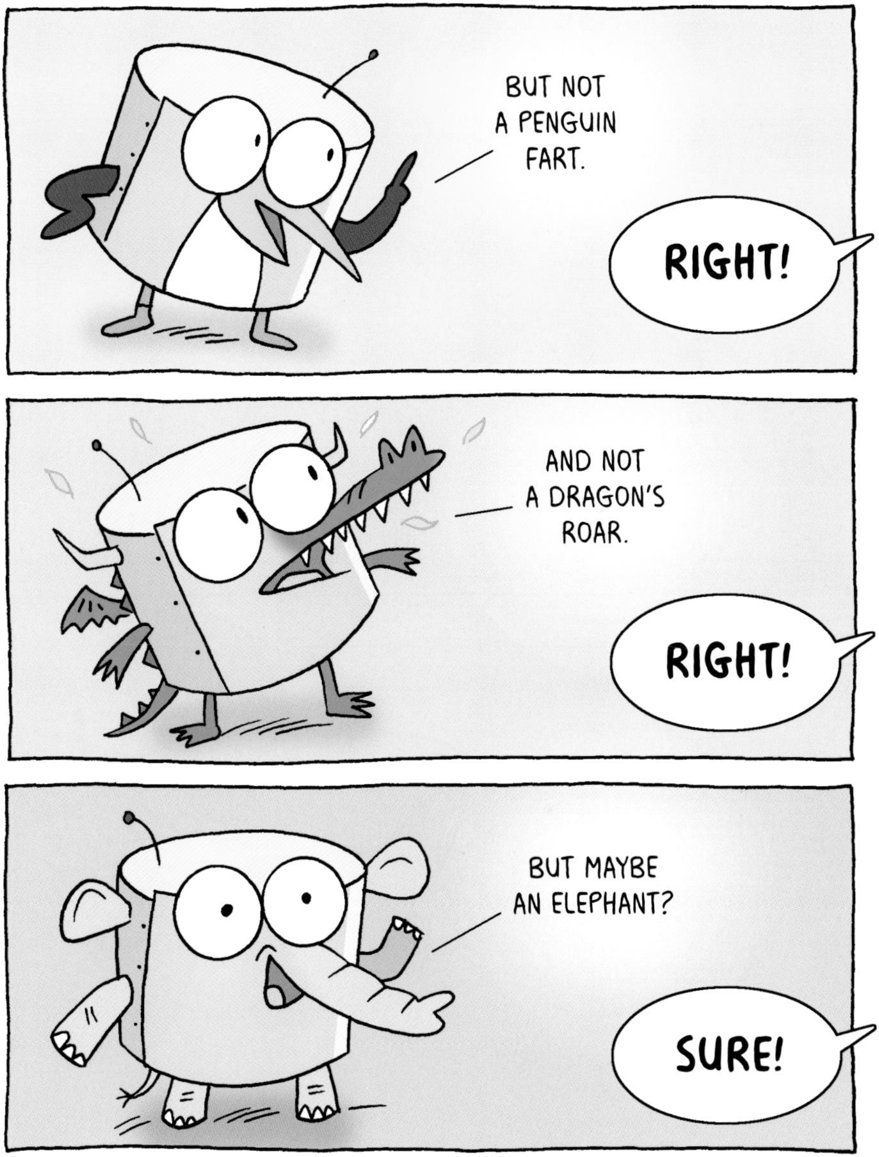
BUT NOT A PENGUIN FART.
RIGHT!
AND NOT A DRAGON'S ROAR.
RIGHT!
BUT MAYBE AN ELEPHANT?
SURE!

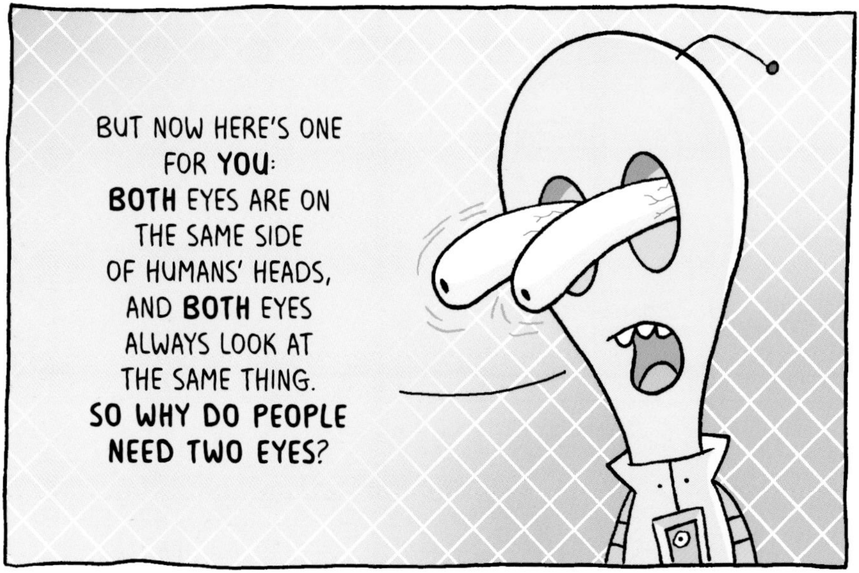
BUT NOW HERE'S ONE FOR **YOU**: **BOTH** EYES ARE ON THE SAME SIDE OF HUMANS' HEADS, AND **BOTH** EYES ALWAYS LOOK AT THE SAME THING. **SO WHY DO PEOPLE NEED TWO EYES?**

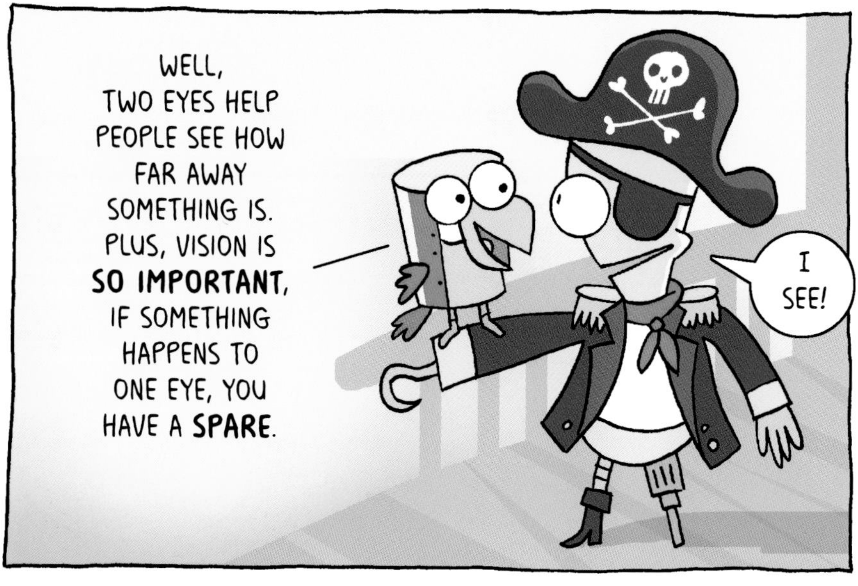
WELL, TWO EYES HELP PEOPLE SEE HOW FAR AWAY SOMETHING IS. PLUS, VISION IS **SO IMPORTANT**, IF SOMETHING HAPPENS TO ONE EYE, YOU HAVE A **SPARE**.
I SEE!

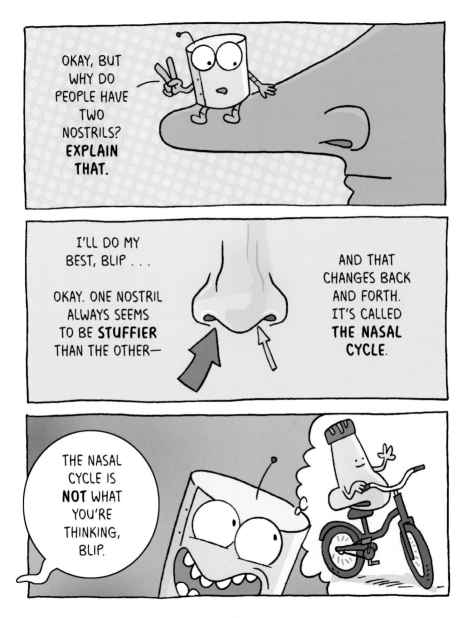

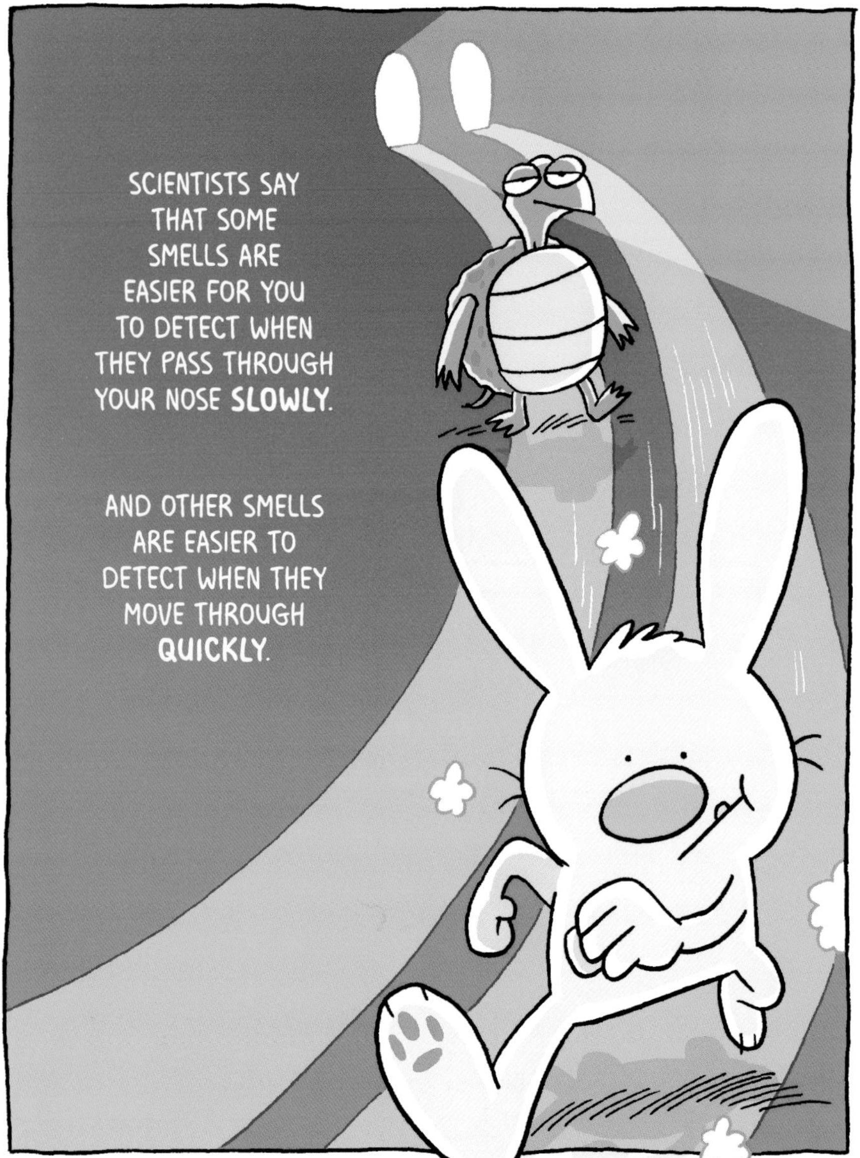
SCIENTISTS SAY THAT SOME SMELLS ARE EASIER FOR YOU TO DETECT WHEN THEY PASS THROUGH YOUR NOSE **SLOWLY**.
AND OTHER SMELLS ARE EASIER TO DETECT WHEN THEY MOVE THROUGH **QUICKLY**.

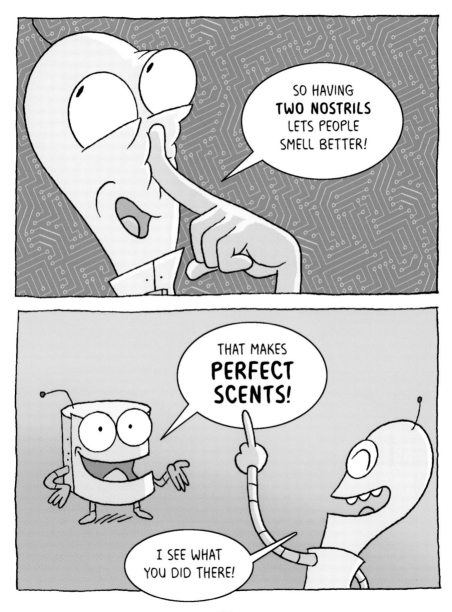

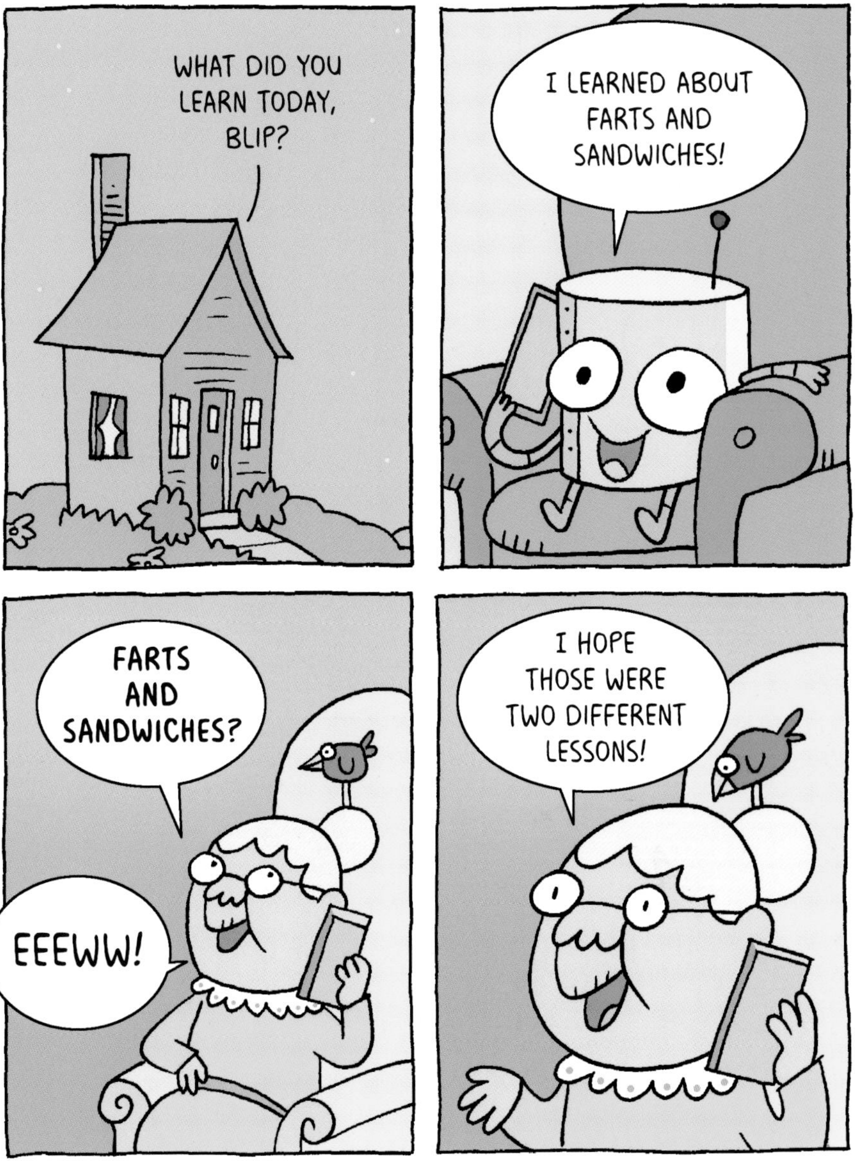
WHAT DID YOU LEARN TODAY, BLIP?
I LEARNED ABOUT FARTS AND SANDWICHES!
FARTS AND SANDWICHES?
EEEWW!
I HOPE THOSE WERE TWO DIFFERENT LESSONS!

YOU KNOW THAT THIS IS **NOTHING** LIKE THE INSIDE OF A PENGUIN,

BUT WE CAN PRETEND IT IS AND SEE IF YOU CAN FIGURE OUT A PATH TO MOVE THE HOT DOG FROM BEGINNING TO END.

ABOUT THE AUTHOR

JIM BENTON IS THE AWARD-WINNING CREATOR OF THE *NEW YORK TIMES* BESTSELLING SERIES DEAR DUMB DIARY, FRANNY K. STEIN, CATWAD, *COMET THE UNSTOPPABLE REINDEER*, AND THE IT'S HAPPY BUNNY BRAND. HIS BOOKS HAVE SOLD MORE THAN FIFTEEN MILLION COPIES IN OVER A DOZEN LANGUAGES AND HAVE GARNERED NUMEROUS HONORS. LIKE JOP AND BLIP, HE ALSO ASKS WEIRD QUESTIONS ALL THE TIME.